Laboratory Investigations
in Human Physiology

Laboratory Investigations in Human Physiology

George K. Russell

Department of Biology, Adelphi University, Garden City, New York

Macmillan Publishing Co., Inc.
New York
Collier Macmillan Publishers
London

Copyright © 1978, George K. Russell

Printed in the United States of America

All rights reserved. No part of this book may be reproduced or transmitted in any form or by any means, electronic or mechanical, including photocopying, recording, or any information storage and retrieval system, without permission in writing from the Publisher.

Macmillan Publishing Co., Inc.
866 Third Avenue, New York, New York

Collier Macmillan Canada, Ltd.

ISBN 0-02-404680-9

Printing: 1 2 3 4 5 6 7 8   Year: 8 9 0 1 2 3 4

ACKNOWLEDGMENTS

The author wishes to express his deep appreciation to the Myrin Institute for Adult Education and to the Doris Duke Foundation, both of New York, for their generous support in the preparation of this manual and to Adelphi University for a sabbatical leave during which time much of this material was compiled. He is particularly grateful to John F. Gardner and to the officers of the Myrin Institute, especially the late Franz E. Winkler, M.D., for advice and encouragement. The idea for this manual emerged from a series of seminars conducted by the Myrin Institute in 1971-72 in which the fundamental role of education in fostering respect for life through direct contact with the living world was emphasized.

The author thanks Thomas E. Fortier for his very able assistance in the preparation of many of the exercises in the manual, Dr. Martha Pavlovich for her work on the exercises dealing with the blood, and W. P. Jennerjahn for the excellent illustrations he prepared for the manual. Many colleagues, too numerous to mention, were also very kind in offering practical advice and assorted pieces of laboratory equipment, as well as in critically evaluating portions of the manuscript. Especial thanks are due to Mrs. Penny Jaworski, Mrs. Shirley Komansky, Regina Quigley, and Barbara Minch for their patient and careful typing of the manuscript.

Figure 1-1, 2-2, and 21-2 are reprinted with permission of Macmillan Publishing Co., Inc., from *Laboratory Manual of Mammalian Physiology* by Barbara A. Shirley. Copyright © 1975 by Barbara A. Shirley.

Figure 1-2 is reproduced in modified form with permission of Educators Publishing Service, Cambridge, Mass., from *Experimental Physiology and Applied Biology* by Tassos P. Filledes. Copyright 1975 by Tassos P. Filledes.

Figure 3-1 is reprinted with permission of Millipore Corporation, Bedford, Mass.

Figure 4-2 is taken from *Review of Physiology,* 3rd ed., by L. L. Langley, copyright © 1971 by McGraw-Hill, Inc. Used with the permission of McGraw-Hill Book Company.

Figures 5-2 and 7-2 are reproduced from H. A. deVries, *Laboratory Experiments in Physiology of Exercise,* Wm. C. Brown Company Publishers, Dubuque, Iowa, 1971, by permission of the author and publisher.

Figure 6-3 is reprinted from *The Cardiac Rhythm, A Systematic Approach to Interpretation* by Raymond E. Phillips and Mary K. Ferney, © 1973 by W. B. Saunders Company, Philadelphia, with permission of the author and publisher.

Figure 7-1 was kindly provided by Quinton Instruments, Seattle, Washington.

The passage on page 118 is quoted from *Textbook of Medical Physiology,* 2nd ed., by A. C. Guyton, © 1961 by W. B. Saunders Company, Philadelphia, with permission of the author and publisher.

Figures 12-1, 12-2, and 13-1 are reprinted with permission of Macmillan Publishing Co., Inc., from *The Human Body: Its Structure and Function,* 3rd ed., by Sigmund Grollman. Copyright © 1974 by Sigmund Grollman.

Figure 13-3 is reprinted with permission of Macmillan Publishing Co., Inc., from *Elementary Physiology and Anatomy,* 3rd ed., by Robert E. Haupt et al. Copyright © 1972 by Robert E. Haupt, Delma E. Harding, Oscar E. Tauber, and Adela S. Elwell.

Figure 13-5 is reprinted from *Physiology and Biophysics,* 19th ed., by T. C. Ruch and H. D. Patton, © 1965 by W. B. Saunders Company, Philadelphia, with permission of the author and publisher.

Figure 14-1 by Biagio J. Melloni is reprinted with permission from *The Human Body: Its Structure and Function,* 3rd ed., by Sigmund Grollman, Macmillan Publishing Co., Inc. Copyright © 1974 by Sigmund Grollman.

Figure 14-2 is reprinted with permission from Harvey Fletcher, *Speech and Hearing in Communication,* 2nd ed., D. Van Nostrand, New York, copyright 1953.

Figure 14-3 is reprinted from *Audiology,* 3rd ed., by H. A. Newby, Prentice-Hall, Inc., Englewood Cliffs, N.J., copyright 1972, by permission of the publisher.

Figures 16-1 and 16-2 were redrawn from illustrations in *Muscles: Testing and Function,* 2nd ed., by H. O. Kendall, F. P. Kendell, and G. E. Wadsworth, © 1971 by The Williams & Wilkins Co., Baltimore, and are included with permission of the authors and publisher.

The passage on page 250 is quoted from *The World of Nigel Hunt* by Nigel Hunt, copyright © 1967 by Garrett Publications, New York, by permission of the publisher.

The Horne-Östberg Morningness/Eveningness Questionnaire is used with the kind permission of Dr. J. A. Horne, Department of Human Sciences, University of Technology, Loughborough, England, and Gordon and Breach Science Publishers Ltd., London, publishers of the *International Journal of Chronobiology.*

Preface

This laboratory manual is intended for a one-semester (or a one-quarter) course in physiology at the undergraduate level. It presumes a knowledge of introductory college biology and introductory college chemistry. Although the manual is designed for a course in human physiology, it can be used equally well in courses of general or mammalian physiology. I have tried to provide representative experiments covering most of the basic topics in physiology, within which the instructor is offered a considerable range of choice. The manual is not designed to accompany any particular textbook of physiology; it may well be used with most of the existing texts.

My principal goal in writing this manual was to provide a series of interesting experiments in basic physiology that make use of the students themselves as experimental subjects, rather than the usual laboratory animals. In my experience, undergraduate students are extremely eager to participate in laboratory studies of this kind. Taught in this way, students have a living encounter with the actual physiological phenomena and are able to relate explanations and mechanisms to a solid base of personal experience. In this connection it is especially interesting to note that experiments with human subjects are playing an increasingly important role in medical school curricula. According to a report presented in *The Physiology Teacher,* a publication of the American Physiological Society,

> There seems to be a trend toward utilizing experiments that can be done on humans and minimizing those done on animals. One reason given is that the student being introduced to bedside teaching early in medical school receives practical demonstration of physiology at the bedside and has much less interest in performing experiments. These labs have offerings of physical diagnosis, pulmonary function tests, electroencephalograms and ECG readings, exercise physiology, special senses and many others which use the students themselves as subjects.[1]

All of the exercises presented are safe, provided that ordinary precautions are observed. I have purposely omitted experiments (e.g., rebreathing from a closed paper bag and stress electrocardiography) which I feel might cause difficulties for an occasional student. The laboratory instructor must assume full responsibility for implementing normal safety procedures, for establishing that students who take part in

---

[1] J. D. Poland, K. E. Guyer, Jr., and H. R. Seibel, Trends in physiology laboratory programs for first year medical students, *Physiol. Teacher,* 4(2):6–8(1975).

these exercise do not have health conditions that could be exacerbated by the tests, and for proper conduct of the experiments. Although no clear-cut guidelines have been prepared by the American Physiological Society for the use of human subjects in undergraduate teaching situations, a statement issued by the American Psychological Society is germane.

> Generally stated, [ethical] principles oblige the investigator to inform participants of all features of the investigation (including risks of physical or mental discomfort) that reasonably might be expected to influence their willingness to participate and to respect their freedom to decline or to discontinue participation at any time . . . . The investigator is also obliged to keep confidential any information obtained about participants during the investigation.[2]

In all cases the principle of "informed consent" must apply, and this too is the responsibility of the instructor.

Although the use of animal experimentation in medical and biological research for which proper justification can be provided is both necessary and important, one can question the need to kill animals in undergraduate teaching situations, especially when valid alternatives are known or could be developed. In my view, undergraduate animal experiments often inflict unnecessary suffering on animals; they also have a hardening and desensitizing effect on the students required to perform them at a time when the development of a sympathetic attitude toward the natural world may be just as important as the teaching of actual scientific knowledge.[3] An imaginative teacher should find ways to experiment on living animals with noninvasive techniques, or to use the students themselves to illustrate the points under consideration. I sincerely hope that many instructors will adopt this approach and help to formulate additional laboratory exercises in physiology along the lines suggested here.

One of the important features included in this manual is the annotated film listing in Appendix C. I have personally reviewed each of the films and videotapes, and can attest to their scientific and educational value. These audiovisual materials can be used to help the students form a deeper connection with the topic under consideration. All too often, with today's one-sided educational methods, the student is steeped in theory at the expense of practical understanding. For example, I once discussed with an undergraduate class the chromosomal basis of Down's syndrome (mongolism), only to find that no single member of the class could have recognized or identified an actual Down's patient. The students were able to provide complex explanations of the chromosomal mechanism underlying this condition, but they had no acquaintance whatsoever with overall symptomology or the personality of a Down's child. This fragmentation of knowledge, I feel, has a great deal to do with the lukewarm interest shown by so many students, and can be countered only by a serious effort on the teacher's part to present students with the whole of the subject at hand. The films and videotapes are provided with this end in mind. The videotapes on sickle cell anemia, for example, treat clinical symptoms and therapeutic measures for this disease and relate both to the underlying cellular and molecular causes. A student who views these tapes and is then able to carry out hemoglobin electrophoresis (Exercise 3) will have a sound grasp of both the human and the molecular aspects of this condition. It has been my experience that increased interest and enhanced student motivation fully justify the utilization of extra class time for these audiovisual materials.

[2]*Ethical Principles in the Conduct of Research with Human Subjects.* American Psychological Association, 1973.
[3]G. K. Russell, Vivisection and the true aims of education in biology, *Am. Biol. Teacher,* 34(5):254–57 (1972).

As an aid to the course instructor and teaching assistants, the amounts of reagents, glassware, and other apparatus requiring advance preparation are included in the lists of materials. Most often the amount is given for each pair of students, but in some instances the amounts are given for "one setup." In addition, the name and address of suppliers of unusual equipment and reagents are provided in notes at the end of the materials lists.

Several of the exercises called for in this manual require reasonably specialized equipment (electrocardiograph, bicycle ergometer, Collins respirometer, etc.), and a department is not likely to possess more than one or two of each of these pieces of apparatus. Class instructors may be able to obtain used equipment from local hospitals and medical centers or from private physicians. One practical way to carry out experiments using equipment that is present in limited supply is to conduct several exercises simultaneously. For example, Exercises 5, 6, and 7 on the heart can be conducted at the same time with a portion of the laboratory section working on each exercise. After the entire class has completed Exercise 4 on the cardiac cycle, one group of students can work on blood pressure (Exercise 5), a second can measure EKG's (Exercise 6), and the third group can evaluate cardiovascular status (Exercise 7). After each section has been completed, the students rotate to the next experiment. In this way four laboratory exercises on the heart can be conducted within about two weekly laboratory periods, and only one EKG machine is required.

Results and Conclusions sheets are provided at the end of each experiment. These can be used as the basis for detailed laboratory reports covering the work of each exercise. *Study Questions* dealing with the content and subject area of each exercise are also provided. In some instances, the study questions go beyond the actual findings of the experiments, and it is hoped that these will help to integrate laboratory material with topics covered in the lecture portion of the course. The citation of pertinent literature at the end of each exercise is intended to supplement the text of the course.

G. K. R.

Contents

Section I — THE BLOOD

1	Some Basic Aspects of the Blood	3
2	White Blood Cells and Disease	17
3	Electrophoresis of Human Hemoglobins	27

Section II — THE HEART

4	The Cardiac Cycle in Man	39
5	Human Blood Pressure	49
6	Electrocardiography	57
7	Cardiovascular Function Tests	67

Section III — RESPIRATION

8	Basic Aspects of Respiration	79
9	Pressure and Volume Relationships in Human Respiration	91
10	Pulmonary Function Tests	107
11	Indirect Measurement of Metabolic Rates	119

Section IV — NERVE AND SENSORY PROCESSES

12	Reflexes in Man and the Nerve Impulse	129
13	Some Elements of Human Vision	139
14	Hearing and Balance	157

Section V — DIGESTION

15	Digestion	173

Section VI — MUSCLES

16	Muscles: Some Elementary Considerations	189
17	Principles of Muscle Physiology	201

Section VII — CELLULAR AND SUBCELLULAR PROCESSES

18	Osmotic Phenomena and Cell Permeability	213
19	Enzymes	225
20	Karyotyping of Human Chromosomes and Amniocentesis	241

Section VIII — RENAL FUNCTION

21	Elementary Urinalysis	259
22	Regulation of Water and Salt Balance by the Kidneys	273
23	Renal Clearance — A Test of Kidney Function	281

Section IX — REGULATORY PROCESSES

24	Glucose Tolerance Test	291
25	Regulation of Body Temperature: The Use of Radiotelemetry	301
26	Buffers and the Constancy of Blood pH	309
27	Circadian Rhythms in Man	317
	Appendix	331

SECTION I

The Blood

EXERCISE 1

Some Basic Aspects of the Blood

OBJECTIVES In this laboratory exercise we shall count two of the cell types found in the blood (erythrocytes, leukocytes) and perform several analytical tests to familiarize ourselves with some basic properties of the blood. The tests to be carried out are (1) hematocrit determination, (2) measurement of hemoglobin concentration, (3) blood typing, and (4) blood clotting time.

The blood constitutes a considerable portion of the fluid content of the human organism. Blood itself is alive. It is subject to the laws of growth, its cellular components undergo highly diversified differentiation, it is extremely active immunologically; in brief, it is a tissue just like the other tissues of the body. It is a remarkable, unusual tissue in that, being a liquid, it lacks a form of its own and must conform to the space it occupies within the cardiovascular system at any given moment. The blood performs a wide variety of essential physiological functions, among which are transporting oxygen and nutrients to the organs of the body, removing wastes, providing a defense against infectious agents and other foreign elements, and serving as a medium for the warmth of the body.

The liquid portion of the blood, the plasma, contains many substances, including soluble proteins, salts, organic nutrients, hormones, antibodies, clotting factors, nitrogenous waste materials, and dissolved gases. The transport and removal of these materials to and from the tissues of the body are essential for cellular life.

The formed elements or blood cells perform several functions in the body. Erythrocytes (red cells) contain the red pigment hemoglobin that functions in the transport of oxygen and carbon dioxide in the blood. Any deficiency in the amount of this essential pigment, due to a reduction either in the total number of erythrocytes or in the actual intracellular concentration of hemoglobin, reduces the oxygen-carrying capacity of the blood. Determination of the number of red cells in a cubic millimeter of blood and the hemoglobin content of the blood are important diagnostic measures.

Leukocytes or white blood cells, many of which move in a typically ameboid fashion, play significant roles in the body's system of defenses against infections and foreign elements. Any decrease in the number or efficacy of these cells results in an enhanced susceptibility to disease. An increase in the number of one or more of the various leukocyte types may indicate the presence of an infectious disease or a parasitic infection, an allergic response, an abnormality in the blood-forming capacity of the body, or other pathological conditions.

ERYTHROCYTE COUNT

The number of erythrocytes in the blood (usually expressed as the number/mm^3) is an important diagnostic measure and gives an indication of the overall health of an individual. Normal adult men have 5.4 ± 0.8 million red cells/mm^3; healthy adult women have 4.8 ± 0.6 million/mm^3.

Materials

sterile finger lancets
sterile cotton
70% alcohol
laboratory Wipettes
RBC pipets (Thoma) with rubber tubes and disposable mouthpieces
hemacytometers (blood-counting chambers)
coverglasses
Hayem's solution (1 bottle per four students)
microscopes

Procedures

Dilution of Blood Sample
1. Cleanse your finger with 70% alcohol, allow it to dry, and puncture it with a sterile lancet. Discard the first drop. Hold your finger vertically with the tip down and allow a large drop to form. Do not squeeze your finger to increase the flow.
2. Draw blood into the capillary bore of the Thoma pipet to the 0.5 mark. Remove excess blood from the outside of the pipet with a laboratory Wipette. The blood level must be maintained at the 0.5 mark by carefully holding the pipet in a horizontal position.
3. Gently suck Hayem's solution into the pipet until the volume of diluted cell suspension reaches the 101 mark.
4. Remove the rubber tubing from the pipet and mix the cell suspension by gently rotating the pipet in the horizontal plane for 2 or 3 minutes. The sample is now diluted, mixed, and ready to count. (By what factor has the blood been diluted?)

Preparation of Hemacytometer and Counting of Cells
1. Place a coverglass on the hemacytometer over the silvered counting area.
2. Discard the first 3 drops in the capillary tip of the blood pipet. Place the tip of the pipet against the edge of the coverglass and allow the cell suspension to seep under the coverglass until the chamber has been filled. Do not add too much liquid. Be certain to study Figure 1-1 for appropriate details of the hemacytometer.
3. Carefully inspect the chamber for signs of flooding or the presence of air bubbles. If there is evidence of either, rinse the hemacytometer and add a new blood sample.

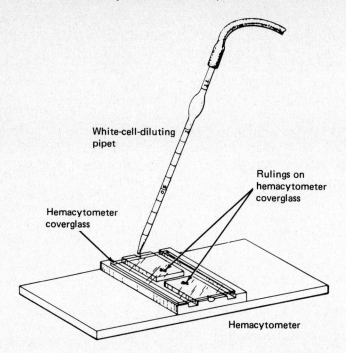

FIGURE 1-1 Technique for applying a diluted blood sample to the hemacytometer.

4. The counting chamber itself is subdivided into nine squares, each of which is 1 mm² in area and 0.1 mm deep. Several of the nine squares are subdivided into 25 smaller squares; each of the 25 is delineated by a double line. We will count the number of cells in one of these squares (area: 0.04 mm², depth: 0.1 mm). See Figure 1-2.

5. Count the number of cells in one square. Calculate your red cell count (as cells/mm³) by multiplying this number times the various dilution factors (×25 to convert to 1 mm²; ×10 to correct for depth; ×200 to correct for pipet dilution). Make four more counts in the same way. Enter your red cell count on the data sheet at the end of this exercise as the average of the five values.

FIGURE 1-2 Hemacytometer chamber for red and white cell counts. Count the white cells in the four large chambers under low magnification; count the red cells in the five small chambers under high magnification.

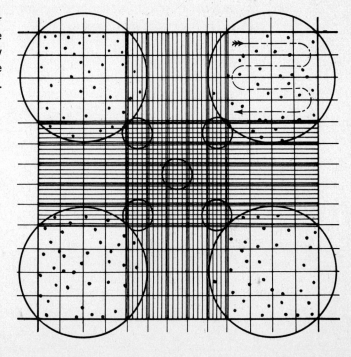

6   Section I]   The Blood

LEUKOCYTE COUNT

The white cells or leukocytes are less numerous than red cells (healthy men have 7000–9000/mm^3; healthy women have 5500–9000/mm^3) and, unlike red cells, are extremely diverse in form and function. In this part of the exercise we shall count the number of white cells in an appropriately diluted blood sample.

Materials

sterile finger lancets
sterile cotton
70% alcohol
WBC diluting pipets with rubber tubing and disposable mouthpieces
hemacytometers
coverglasses
Turk's solution (1 bottle per four students)
microscopes

Procedures

Dilution of Blood Sample
1. Cleanse the tip of your finger with 70% alcohol, allow it to dry and lance it. Discard the first drop of blood and allow a second to form.
2. Using the white cell pipet, draw the blood up to the 0.5 mark, and carefully add Turk's solution to the 11.0 mark. Mix as before. (Turk's solution stains the leukocytes and causes the red cells to lyze.)

Counting of Cells
1. Discard the first 4 drops from the pipet and fill the hemacytometer as described above for the erythrocyte count.
2. With the low power objective of the microscope, locate the four large squares as shown in Figure 1-2.
3. Switch to high power, focus on the upper right square, and count the number of leukocytes by following the pattern indicated in Figure 1-2. Count the numbers in all four large squares and average the results.
4. Calculate the white cell concentration (cells/mm^3) by correcting the average count ($\times 10$ to correct for depth; $\times 20$ to correct for pipet dilution). What is your leukocyte count? Enter the value on the data sheet.

HEMATOCRIT DETERMINATION

This test specifically measures the relative volumes of plasma and cellular elements in the blood. A small sample of whole blood is centrifuged in a special capillary tube under standardized conditions. After centrifugation the hematocrit value (often called the packed cell volume) is determined as the ratio of packed red cells to the volume of the original sample, multiplied by 100.

Exercise 1] Some Basic Aspects of the Blood

The average hematocrit value for men is about 46% with a range of 43–49%; women show a somewhat lower mean value of 41% with a range of 37–45%. The test gives very low values in certain forms of anemia and very high values in the disease polycythemia.

Materials

sterile finger lancets
sterile cotton
70% alcohol
laboratory Wipettes
heparinized capillary tubes
Seal-Ease (clay compound)
hematocrit centrifuge
hematocrit reader

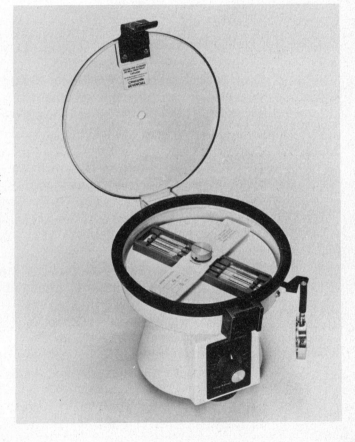

FIGURE 1-3 The Clay Adams Readacrit centrifuge for direct determination of hematocrit values. [*Courtesy of Clay Adams.*]

Procedures

1. Cleanse the tip of your finger with 70% alcohol, allow it to dry, and lance it as described above. Discard the first drop of blood and allow a second to form.
2. Touch a heparinized capillary tube to the drop and allow the blood to fill it.
3. Seal one end of the capillary with clay compound (Seal-Ease).
4. Place the tube in a hematocrit centrifuge with the clay-plugged end directed outward. Secure the top of the centrifuge and spin for 5 minutes. Figure 1-3 illustrates the Clay Adams Readacrit centrifuge.

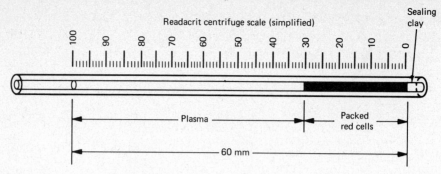

FIGURE 1-4 Hematocrit capillary tube containing blood sample after centrifugation. [*Courtesy of Clay Adams.*]

5. Remove the tube from the centrifuge and determine the hematocrit value as the ratio of packed cells to the total blood volume (Figure 1-4). Enter your value on the data sheet.

DETERMINATION OF HEMOGLOBIN CONCENTRATION

Hemoglobin, a protein found abundantly within red blood cells, is chiefly concerned with transport of oxygen to the tissues of the body. Oxygen combines with hemoglobin in a freely reversible fashion as follows:

$$Hb + 4O_2 \rightleftharpoons Hb4O_2$$

Arterial blood contains primarily oxygenated red hemoglobin; venous blood contains deoxygenated hemoglobin which is reddish blue in color. (Do these statements hold true for the pulmonary artery and vein?) Carbon monoxide has a very high affinity for hemoglobin (210 times as great as oxygen) and binds at the oxygen-carrying sites. Cyanide also binds to hemoglobin to form a stable complex, and this reaction can be used to measure hemoglobin concentrations.

Quite obviously, the oxygen-carrying capacity of the blood depends upon the hemoglobin content. Each gram of hemoglobin, when fully saturated with oxygen, transports 1.33 ml of oxygen. In a healthy adult man the hemoglobin content is approximately 16 grams/100 ml of blood, in a healthy adult woman approximately 14 grams/100 ml of blood.

In this part of the exercise we shall measure our hemoglobin content and determine the oxygen-carrying capacity of our blood.

Materials

sterile finger lancets
sterile cotton
70% alcohol
Tallquist blotting papers and color comparison chart
test tubes (8 per two students)
test tube racks (1 per two students)
cyanmethemoglobin standard solution (hemoglobin = 80 mg/100 ml) (20 ml per pair of students)

Exercise 1] Some Basic Aspects of the Blood

cyanmethemoglobin reagent solution (25 ml per two students)
20 µl Sahli blood pipets and rubber tubes with disposable mouthpieces
rubber propipetting bulbs (1 per four to eight students)
5.0 ml pipets (2 per two students)
Bausch & Lomb Spectronic 20 spectrophotometers (or other suitable instruments) and spectrophotometer tubes

Procedures

Tallquist Method
1. Disinfect your finger with 70% alcohol, allow it to dry, and puncture it with a sterile lancet as described above.
2. Apply a drop of blood to a piece of blotting paper and read the hemoglobin concentration by comparing the paper with the Tallquist standards. What is the hemoglobin concentration of your blood in grams per 100 ml? Record the value on the data sheet.

Cyanmethemoglobin Method
1. Set up the materials for a standard curve by pipetting various amounts of the cyanmethemoglobin standard solution and reagent into a series of tubes, as follows:

Tube #	Cyanmethemoglobin Standard Solution	Reagent
1	1.0 ml	4.0 ml
2	2.0 ml	3.0 ml
3	3.0 ml	2.0 ml
4	4.0 ml	1.0 ml
5	5.0 ml	0
6	0	5.0 ml

2. Disinfect your finger, puncture it with a sterile lancet and collect exactly 0.02 ml of blood with a Sahli blood pipet. Quickly add the sample to 5.0 ml of the reagent in a test tube and mix thoroughly. Allow the tubes to stand for at least 10 minutes.
3. Read the absorbance of the standards and your blood sample in a spectrophotometer at wavelength 540 nm. Use tube #6 as the blank for all spectrophotometer readings.
4. Plot a standard curve as absorbance versus hemoglobin concentration and calculate the hemoglobin concentration of your blood. Enter the standard curve and your hemoglobin value on the data sheet at the end of the exercise.

BLOOD TYPING

Blood typing is based upon the presence of certain antigenic substances in the blood, many of which are localized on the red cell membrane. The best-known antigens comprise the ABO series, originally described in 1900 by Dr. Karl Landsteiner.

TABLE 1-1 Antigens and Antibody Content of the Blood Types of the ABO Series

		Blood Contains	
Genotype	Blood Type	Cellular Antigens	Plasma Antibodies
ii	O	none	anti-A and anti-B
$I^A I^A$ or $I^A i$	A	A	anti-B
$I^B I^B$ or $I^B i$	B	B	anti-A
$I^A I^B$	AB	A and B	none

An individual specifically inherits the capacity to produce only the A antigen, only the B antigen, or both (AB). A person producing neither is type O. Individuals also have specific antibodies (agglutinins) in their blood. Table 1-1 summarizes this information. (Recall that genetic factors exist in pairs, each of which is associated with one of a pair of homologous chromosomes.) Table 1-2 illustrates the relative frequencies of the ABO antigens in selected human populations.

Type A individuals produce the A antigen and the anti-B agglutinin. What would happen if a type A person's blood was transferred into a type B person (with B antigen and anti-A agglutinin)? What would happen if the transfusion was done in the reverse manner (i.e., type B blood into an A recipient)? What is a universal donor? What is a universal recipient?

Another substance found in the blood is the Rh antigen. Individuals who produce this are termed Rh^+; nonproducers are Rh^-. When an Rh^- mother gives birth to an Rh^+ child, it may happen that the Rh^+ antigen passes into the mother's blood and sensitizes her (i.e., causes her to produce antibodies against the Rh^+ antigen). This poses no danger to the mother or to the first child, but a subsequent Rh^+ fetus may have its red cells agglutinated and suffer the condition known as erythroblastosis fetalis. For this reason, Rh evaluation is an important test for all prospective parents.

Approximately 85% of white women in the United States are Rh^+; 15% are Rh^-. The incidence of the Rh^+ condition is higher in blacks and orientals.

Materials

anti-A, anti-B, and anti-Rh blood sera
slide warmer
sterile finger lancets
sterile cotton
70% alcohol
microscope slides (2 per student)
wax marking pencils
thin wooden sticks

Procedures

1. Divide the microscope slides into three equal squares with the wax pencil. Label one end A, the other end B.

Exercise 1] Some Basic Aspects of the Blood

TABLE 1-2 Frequencies of ABO Blood Groups in Selected Human Populations[a]

Population	Frequency (%)			
	O	A	B	AB
U.S. whites	45	41	10	4
U.S. blacks	47	28	20	5
African pygmies	31	30	29	10
African bushmen	56	34	8	2
Australian aborigines	31	66	0	0
Pure Peruvian Indians	100	0	0	0
Tuamotuans of Polynesia	48	52	0	0

[a]Reprinted from *Biological Science,* 2nd ed., by William T. Keeton. Illustrated by Paula DiSanto Bensadoun. By permission of W. W. Norton & Company, Inc. Copyright © 1967, 1972 by W. W. Norton & Company, Inc.

2. Carefully place 1 drop of anti-A serum in square A, and 1 drop of anti-B serum in square B. Place 1 drop of anti-Rh serum on a second slide.
3. Disinfect your finger and puncture it with a sterile lancet. Allow 1 or 2 drops of blood to fall onto each of the three anti-sera. Do not touch the anti-sera with your finger.
4. Mix the blood and anti-sera with the small wooden sticks and observe the results. The Rh test slide is to be heated to 45°C on the slide warmer. Agglutination is indicated by a "grainy" appearance of the blood sample and should occur within 2 minutes. If you do not get positive results with your blood, examine the slides of other students to see what an agglutination reaction looks like. If time permits, try cross-matching your blood with others in the class.
5. What is your blood type? Do your class results approximate the population as a whole? Who in the class is a universal donor? Who is a universal recipient? Which female students are Rh$^-$?

CLOTTING

The clotting of the blood is a highly complex process involving at least thirty separate factors in the blood. The basic mechanism involves the formation of a substance called prothrombin activator in response to the rupture of a blood vessel or to a chemical alteration of the blood itself. This activator catalyzes the conversion of prothrombin to thrombin, and the thrombin acts as an enzyme to convert the soluble protein fibrinogen to fibrin threads. These threads form a fibrous network within which red cells and plasma are enmeshed to form the gelatinous substance of the clot. Fibrin is insoluble and filamentous.

In this part of the exercise we shall observe the formation of fibrin strands and measure the clotting time of the blood.

Materials

sterile finger lancets
sterile cotton
70% alcohol

clean microscope slides and coverglasses
microscopes
unheparinized capillary tubes
small mechanical file
methyl violet stain (1 bottle per four students)

Procedures

Fibrin Formation
1. Disinfect your finger, allow it to dry, and lance it.
2. Place a drop of blood on a clean slide and cover it with a coverglass.
3. View the slide with the low power objective of your microscope. It may be helpful to add a drop of methyl violet solution to the edge of the coverglass.
4. Observe the formation of fibrin. Make a clear drawing of what you see.

Clotting Time
1. Again disinfect and lance your finger. Note the exact time at which the flow of blood commences.
2. Rapidly draw blood into a capillary tube by holding the tube directly in the drop of blood in a horizontal position.
3. At 30 second intervals break off small portions (about 0.5 cm) of the capillary and see if clotting has occurred. This procedure is facilitated very greatly if the capillary tube has been previously scratched with a small file at 0.5 cm intervals.
4. Separate the broken ends slowly and gently while looking for coagulation. Clotting has occurred when threads of fibrin span the gap between the broken ends of the tube.
5. Record your clotting time.

STUDY QUESTIONS

1. What would happen to your erythrocyte count if you moved from sea level to elevation 10,000 feet and lived there for a period of time? If you suffered from an iron deficiency? If you trained for athletic competition by running 5 miles a day for several months? If you had a chronic respiratory condition such as emphysema? What are the specific mechanisms by which the body regulates the number of erythrocytes in the blood?

2. What would you expect the hematocrit reading to be in each of the circumstances described above? Explain your answers.

3. The molecular weight of hemoglobin is 64,000 daltons. Use the figure that you derived from the amount of hemoglobin/red blood cell to determine the number of hemoglobin molecules in each of your erythrocytes.

4. Complete the following chart by filling in the blanks:

Blood Type	Can Donate to	Can Receive from
O		
A		
B		
AB		

5. Define or characterize anemia, polycythemia, leukopenia, leukemia, leukocytosis.

6. An Rh^- woman may bear an Rh^+ child if the father is Rh^+. Why is there no difficulty with the first Rh^+ birth? What measures can be taken to prevent erythroblastosis fetalis in any subsequent pregnancies?

7. Hemophilia is an inherited disease that usually affects only men, although very rare female hemophiliacs are known. Women may serve as carriers of the disease, however, because the condition is inherited as a sex-linked recessive. Queen Victoria is perhaps the best-known example of a female carrier. What is the molecular basis of hemophilia? Based on what you know about its molecular basis, why would you expect hemophilia to be recessive rather than dominant? Can you think of a way to test women to determine if they are heterozygous (carriers) for this trait?

REFERENCES

Adamson, J. W., and C. A. Finch. 1975. Hemoglobin function, oxygen affinity, and erythropoietin. *Ann. Rev. of Physiol., 37*:351.

Arnone, A. 1974. Mechanism of action of hemoglobin. *Ann. Rev. of Med., 25*:123.

Frei, E., III, and E. J. Freireich. 1964. Leukemia. *Sci. Amer., 210*(5):88.

Gordon, A. S. 1973. Erythropoietin. *Vitamins and Hormones, 31*:106.

Harker, L. 1974. Control of platelet production. *Ann. Rev. of Med., 25*:383.

Laki, K. 1962. The clotting of fibrinogen. *Sci. Amer., 206*(3):60.

McConnell, R. B., and J. C. Woodrow. 1974. Immunoprevention of Rh hemolytic disease of the newborn. *Ann. Rev. of Med., 25*:165.

Ponder, E. 1957. The red blood cells. *Sci. Amer., 196*(1):95.

Ross, R. 1968. Fibroblasts and wound repair. *Biol. Rev., 43*(1):51.

Seegers, W. H. 1969. Blood clotting mechanisms: three basic reactions. *Ann. Rev. of Physiol., 31*:269.

Spiers, R. S. 1964. How cells attack antigens. *Sci. Amer., 210*(2):58.

Zucker, M. B. 1961. Blood platelets. *Sci. Amer., 204*(2):58.

EXERCISE 1

Name _____

Laboratory Section _____   Date _____

RESULTS AND CONCLUSIONS

Erythrocyte Count

Erythrocyte count = _____ cells/mm^3

Leukocyte Count

Leukocyte count = _____ cells/mm^3

Hematocrit Determination

Height of total column (cells plus plasma) = _____ mm

Height of packed cell column = _____ mm

Hematocrit value (packed cell volume as percent of total volume) = _____ %

Hemoglobin Determinations

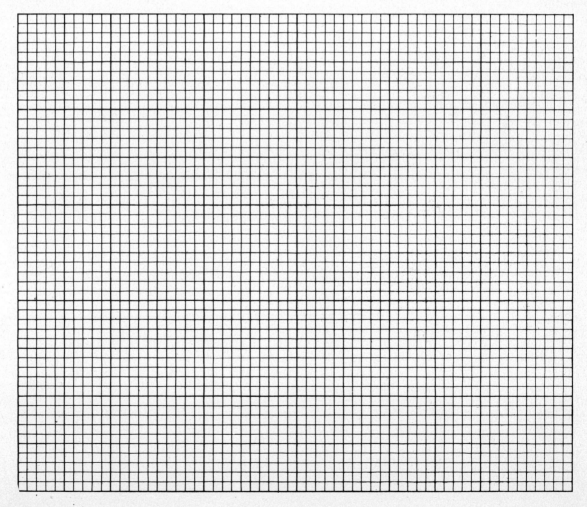

Exercise 1] Results and Conclusions

Hemoglobin concentration (Tallquist) = _____ grams/100 ml

Hemoglobin concentration (cyanmethemoglobin) = _____ grams/100 ml

Hemoglobin count = _____ %

> Use 16 grams/100 ml = 100% for men
> Use 14 grams/100 ml = 100% for women

Hemoglobin content/cell = _____ $\times 10^{-11}$ gram/cell

Oxygen-carrying capacity of blood = _____ ml O_2/100 ml blood

Blood Typing

Reaction with anti-A agglutinin_____

Reaction with anti-B agglutinin_____

Reaction with anti-Rh agglutinin_____

Blood type = _____

Clotting

Clotting time = _____ min

EXERCISE 2

White Blood Cells and Disease

OBJECTIVES In this exercise we shall prepare a blood smear and perform a differential white blood cell count. The different types of leukocytes and their relative percentages are to be determined. Prepared microscope slides showing various pathological conditions will be provided for detailed study and comparison with normal blood. Finally, we shall view the excellent biomedical film *The Embattled Cell,* which illustrates the various defense mechanisms the body uses in combating cancer.

White blood cells, unlike red cells, exhibit a wide diversity of form and function. Most cytologists distinguish five basic types of leukocytes in normal blood. Each type shows a characteristic cellular morphology, as illustrated in Figure 2-1, and performs essential functions in the body's system of defenses.

Type	*Percent of Total Leukocytes*
Neutrophils	54–62
Lymphocytes	25–33
Monocytes	3–7
Eosinophils	1–3
Basophils	0.5

Neutrophils, eosinophils, and basophils all contain a highly granular cytoplasm and are collectively known as granulocytes. These cells are also called polymorphonucleocytes ("polymorphs") because of the complex appearance of their nuclei. Together they constitute a substantial fraction of the total white cell population of the body.

Neutrophils

Neutrophils are the predominant form of leukocyte in the blood. These cells typically show small rose-pink or purple granules in the cytoplasm and a complex purple nucleus (two to five lobes) when treated with Wright's stain. They are 10–12 microns in diameter. Neutrophils are often called phagocytes because they engulf,

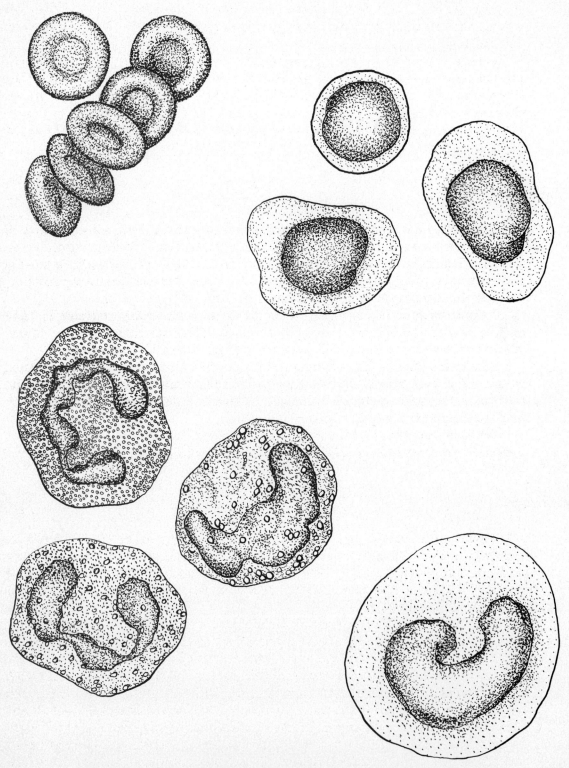

FIGURE 2-1 The different cell types found in human blood. *Upper left:* Erythrocytes. *Upper right:* Large and small lymphocytes. *Lower left:* Three types of granulocytes. *Lower right:* Monocyte.

ingest, and destroy bacteria, necrotic tissue, and other elements foreign to the body. They are found in vast numbers at the site of an infection, to which they are drawn by chemical attraction, and as phagocytes they constitute the first line of defense against invading bacteria. The granular nature of their cytoplasm is due in part to the presence of lysosomal particles containing highly active enzymes that aid in the breakdown of ingested material.

Neutrophils originate in the bone marrow. They are unable to divide mitotically and live only a short time. Those that enter the circulation and remain there for their entire lifespan live about 30 hours. Many leave the bloodstream by squeezing through the capillary walls by a process termed diapedesis. Once a neutrophil leaves the circulatory system, it rarely returns; it moves into the tissues and organs of the body in ameboid fashion where it assists in combating infection.

Lymphocytes

When treated with Wright's stain, lymphocytes can be recognized by pale to dark blue cytoplasm surrounding a round, oval, or slightly indented nucleus. A small lymphocyte (7 microns in diameter) has a very narrow band of cytoplasm around a prominent nucleus. A large lymphocyte (10 microns in diameter) has an oval-shaped, indented nucleus surrounded by abundant cytoplasm.

Cellular immunity is mediated by small T-lymphocytes that originate in bone marrow and are processed by the thymus gland. These are responsible for hypersensitivity reactions, rejection of transplants, and graft-versus-host reactions.

Humoral immunity, the formation of circulating antibodies in the blood, is based on the activity of plasma cells. B-lymphocytes differentiate into plasma cells and establish residency in germinal centers in the lymph nodes. Plasma cells do not ordinarily circulate freely in the blood.

Detailed treatment of the complex functioning of lymphocytes can be found in standard texts of physiology and immunology.

Monocytes

Monocytes are the largest white cells in the body (approximately 15 microns in diameter), and show a large horseshoe- or kidney-shaped nucleus that stains deep blue or purple with Wright's stain. They show abundant blue-gray cytoplasm containing extremely small clear blue granules that can be seen only under high magnification in a well-prepared blood smear.

The monocytes, like neutrophilic leukocytes, are actively phagocytic and are known to invade areas of inflammation where they remove bacteria and cellular debris by phagocytosis. Upon entering an area where infection or wounding has occurred, monocytes are transformed into very large, active cells called macrophages. Although their detailed structure varies somewhat in different regions of the body, macrophages all contain numerous cytoplasmic granules and can ingest almost any kind of foreign material. Tissue macrophages divide mitotically, but cell division is not the only means by which they replenish their numbers. In response to the presence of microorganisms or other foreign elements in the body, monocytes are transformed into macrophages indistinguishable from those ordinarily found in the tissues.

Eosinophils

Eosinophils are granulocytic leukocytes comprising 1-3% of the total white cell count. They react with Wright's stain to show extremely coarse red-orange granules in the cytoplasm and a blue-purple bilobed nucleus. The cellular diameter is about 13 microns.

The function of eosinophils is largely unknown, although several suggestions have been advanced. They are weakly phagocytic and increase markedly in the blood when foreign proteins are present. Several authorities feel that eosinophils phagocytize antigen-antibody complexes. Also, the level of circulating eosinophils is often elevated in patients with strong allergic conditions.

Basophils

Basophils constitute less than 1% of the total leukocyte count. When treated with Wright's stain, they show large red-purple cytoplasmic granules and a bilobed blue-black nucleus. The cell size is approximately 14 microns in diameter.

Basophils are known to contain the powerful anticoagulant heparin; they may function to prevent intravascular clotting, although experimental evidence to support this view is weak. Basophils are phagocytic, but may not function significantly in this way.

DIFFERENTIAL WHITE BLOOD CELL COUNT

In this part of the exercise we shall perform a differential white blood cell count and examine prepared slides of various cellular pathologies in which the white cell count is abnormal.

Materials

clean microscope slides
coverglasses
sterile disposable lancets
sterile cotton
70% alcohol
Wright's stain (1 small dropping bottle per four students)
phosphate buffer solution: 1.63 grams KH_2PO_4 + 3.2 grams $NaHPO_4$ in 1 liter of
    water, pH 6.4-6.8 (1 small dropping bottle per four students)
staining trays with racks (1 per two students)
microscopes
immersion oil
colored drawing pencils
prepared slides of pathological conditions (see Note)

Note: An assortment of prepared slides—infectious mononucleosis, various leukemias and leukopenias, and others—is available from biological supply companies (e.g., Carolina Biological Supply Co., Wards).

Procedures

1. Disinfect your finger and lance it. Place one drop of blood on a slide, about 1 inch from the end, and place the slide on the laboratory bench in front of yourself.
2. With the thumb and forefinger of your left hand, hold the edge of the slide that is away from the drop of blood. With your right hand, pick up another slide to use as the "spreader."
3. Holding the spreader between the thumb and forefinger, place its leading edge in front of the drop of blood at a 30° angle. The spreader must be maintained at this angle throughout the procedure. Draw the spreader toward the blood and as it contacts the drop allow the blood to fan out to the edges of the spreader.
4. Quickly draw the spreader along the length of the slide to produce a "blood smear." The spreader must be held at the 30° angle throughout the spreading procedure, and it should be pressed firmly against the slide. Note that the blood is to be pulled or drawn across the slide, not pushed. Figure 2-2 illustrates the correct technique.
5. Allow the slide to dry in the air.
6. Place the slide on the rack in the staining tray. Cover the dried blood smear with 50 drops of Wright's stain, *evenly distributed over the entire slide.* Let the slide stand for 2 minutes.
7. Add 25 drops of phosphate buffer solution, and mix it with the stain by blowing from a point about 4 inches directly above the slide. Do not blow the fluids off the slide. When the solutions are well mixed, a shiny scum can be seen floating on top of the solution. Allow the buffer and stain to set for 5 minutes.
8. Wash the slide thoroughly in a beaker of water or under the tap, and shake off the excess water. Stand the slide on end and allow it to dry.

FIGURE 2-2 Preparation of a blood smear.

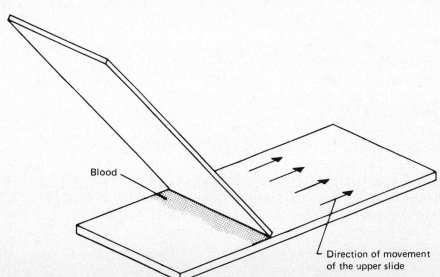

9. Examine the slide under oil immersion to observe the various types of leukocytes. View the slide by examining its surface in a systematic fashion. Count and classify 100 leukocytes, and calculate the relative percentage of each type. The various types are shown in Figure 2-1. Identify each and color it according to what you see in your own preparation.
10. Examine the prepared slides of various pathological conditions and note carefully the number and type of white cells. Which cell type predominates in each case?

THE EMBATTLED CELL

As we have seen, cytological study of many cell and tissue types is often carried out with fixed, stained preparations of dead tissue. Studies of this kind have most certainly yielded important results, but there are inherent methodological limitations in much of this work. Accordingly, many researchers are turning to the study of actual living cells through the use of tissue culture techniques. Portions of living tissue are grown outside the organism in suitable nutrient fluids, and detailed cytological studies, often using time-lapse motion picture photography, are performed.

The American Cancer Society has produced an important scientific film, *The Embattled Cell,* * which shows time-lapse sequences with phase-contrast microscopy of human cancer cells grown in tissue culture and the roles that various types of leukocytes play in defending the body against cancer. There is general consensus among medical researchers that cancer continually arises spontaneously in the human body and that the normal, healthy human organism has the capacity to resist and destroy these cells. The development of cancer is due as much to the incapacity of the human organism to resist as it is to the presence of a virus or carcinogenic agent. A recent statistical study indicates, for example, that kidney transplant patients (all of whom have decreased resistance because they receive immunosuppressive drugs to prevent rejection of their new organs) have eighty times as great a chance of developing malignancy as healthy individuals not under immunosuppression [Penn and Starzl, 1972].

We shall view *The Embattled Cell* and study the roles of the various types of leukocytes in defending the body against disease. Two important questions to keep in mind during the film are

(1) What are the many abnormal properties of cancer cells that are illustrated in the film?
(2) What are the many different lines of resistance, including various types of leukocytes, that defend the body?

A thorough class discussion of both questions following the film will be very helpful.

STUDY QUESTIONS

1. Certain congenital diseases are known in which the immune responses of the body are deficient. These include DiGeorge's syndrome (congenital aplasia of the thymus), agammaglobulinemia (inherited inability to form gamma globulins), and

---

*The Embattled Cell, available on free loan from your local chapter of the American Cancer Society, is a truly excellent instructional film (see page 336).

Exercise 2] White Blood Cells and Disease

severe combined immune deficiency. How does each of these conditions affect the immune responses of the individual? Which lymphocytes are affected in each case?

2. The first heart transplant patient in South Africa did not succumb to rejection of his new heart; rather, his death was attributed to bacterial pneumonia. Can you propose a reasonable explanation for this finding?

3. The cellular changes that you noted in your study of prepared slides of various pathologies can be said to characterize each of these diseases. Are the cellular changes *symptoms* or *causes* in each case?

4. Several different forms of leukemia are known. Identify some of these and characterize the cellular changes that occur in each. How are these changes thought to originate?

5. Infectious disease is usually attributed to an encounter with a pathogenic microbe and the proliferation of this organism within the body. However, we are constantly exposed to microbes without necessarily becoming ill, and the causative microbes of several diseases, such as herpes simplex virus (the cause of cold sores), pneumonia bacteria, and possibly certain cancer viruses, may be present in the body all the time. It follows, then, that the cause of disease is not so much the proliferation of the microbe as it is an alteration of the body's system of defenses. The breakdown of resistance, however, must also be seen as symptomatic. Characterize some of the conditions that might lead to altered resistance in the human body. (You will find *Man's Search for Meaning* [Frankl, 1976] a significant contribution to your study.)

REFERENCES

Cooper, M. D., and A. R. Lawton III. 1974. Development of the immune system. *Sci. Amer., 231*(5):58.

Frankl, V. 1976. *Man's Search for Meaning.* Pocket Books, New York.

Greaves, M. F., J. J. T. Owen, and M. C. Raff. 1974. *T and B Lymphocytes: Origins and Roles in Immune Responses.* American Elsevier, New York.

Jerne, N. J. 1973. The immune system. *Sci. Amer., 229*(1):52.

Lerner, R. A., and F. J. Dixon. 1973. The human lymphocyte as an experimental animal. *Sci. Amer., 228*(6):82.

Penn, I., and T. E. Starzl. 1972. Malignant tumors arising *de novo* in immunosuppressive organ transplant patients. *Transplantation, 14*:407.

Rosen, F. S. 1975. *Immunogenetics and Immunodeficiency,* Chapter 6, "Immunodeficiency." United Park Press, Baltimore.

Ross, R. 1969. Wound healing. *Sci. Amer., 220*(6):40.

Watson, J. D. 1976. *The Molecular Biology of the Gene,* 3rd ed., Chapter 20, "The Viral Origins of Cancer." Benjamin, Menlo Park, CA.

EXERCISE 2

Name _____

Laboratory Section _____ Date _____

RESULTS AND CONCLUSIONS

Differential White Cell Count

	Neutrophils	Eosinophils	Basophils	Monocytes	Lymphocytes
Tally marks					
Totals					
Relative percentages					
Normal values	54–62%	1–3%	0.5%	3–7%	25–33%

Pathological Conditions

Describe in concise and accurate manner the various cellular pathologies you observed in the prepared slides. What cell type(s) predominates in each instance?

The Embattled Cell

Tabulate the various abnormal properties of cancer cells shown in the film, that distinguish cancer cells from normal cells.

Exercise 2] *Results and Conclusions*

Several of the body's defense mechanisms are illustrated in *The Embattled Cell*. For each of the defenses listed below, summarize the findings shown in the film, and discuss how each process protects or defends the body against cancer, emphysema, and other serious conditions. Try to include specific details from the film.

(1) Ciliated cells lining the respiratory passages.

(2) Mucus-producing cells.

(3) Lymphocytes (various types).

(4) Macrophages.

(5) Mast cells.

EXERCISE 3

Electrophoresis of Human Hemoglobins

OBJECTIVES In this exercise we shall study normal erythrocytes and red cells from individuals with sickle cell trait. A phosphate solubility test will be performed on both samples, and sickle cell crisis will be simulated in normal red cells with a chemical sickling agent. Electrophoresis on cellulose acetate strips will be used to separate hemoglobin S from hemoglobin A.

Electrophoresis is a widely used analytical and diagnostic technique employed in medical and biological research to study serum proteins, multiple forms of enzymes, glycoproteins, hemoglobins, and many other biological molecules. In this exercise we shall prepare hemoglobin from our own blood and from sickle cell trait blood and examine these molecules using electrophoresis on cellulose acetate strips.

Sickle cell disease was first described in 1910 by an American physician, Dr. James B. Herrick. It is found predominantly in black people and is inherited as an autosomal recessive gene. Individuals who are homozygous for the recessive allele are seriously afflicted with a wide variety of attendant symptoms. The clinical symptoms vary considerably in different patients, but include chronic anemia, jaundice, decreased physical capability, enhanced susceptibility to infections (especially in the lungs), retardation of growth, fibrosis of the spleen, and malfunction of the heart, kidneys, and lungs. Under low oxygen tension, the red blood cells of these individuals assume a crescent or sickle shape and tend to clump in small blood vessels, bringing about local circulatory failure. The serious consequences of sickle cell disease are traceable to this deficiency of blood supply to the tissues, and to the rapid destruction of sickle cells by the body.

The diagnosis of sickle cell anemia is based on three separate criteria:

(1) The painful episodes or "crises" that afflict severely ill patients.
(2) The sickling phenomenon that occurs when erythrocytes are made anaerobic.
(3) The presence of abnormal hemoglobin in the blood.

People who are heterozygous for the condition (carry the sickle cell trait) are quite normal, although they may experience a rare sickle cell crisis under conditions of extreme physical duress. Such individuals are warned against flying in unpressurized airplanes, sky diving, and underwater activities, all circumstances in which they may experience difficulties. At least one professional athlete is known to be a carrier of the trait, however, and the heterozygous condition is ordinarily not severe. Sickle cell trait is found in about 10% of the blacks in the United States, while sickle cell disease itself occurs in 1 or 2%.

In 1949 Linus Pauling and several collaborators demonstrated, using the technique of electrophoresis, that a heterozygous individual produces two forms of hemoglobin, one being normal and the other identical to the abnormal hemoglobin found in sickle cell patients homozygous for the mutant gene. Subsequent biochemical research has provided clear understanding of the exact nature of the defect. A single amino acid substitution in the sixth position of one of the two types of polypeptide chains of the hemoglobin molecule leads to the sickling of erythrocytes containing this molecule and produces all of the consequent effects described above.

These biochemical studies of this kind have been extremely valuable to students of gene functioning at the molecular level, but have not provided a sound rationale for therapy. Current medical research is directed toward an attempt to alleviate the symptoms of sickle cell disease by preventing sickling through the use of chemical agents such as potassium cyanate [Cerami and Peterson, 1975].

Materials

70% alcohol
sterile finger lancets
sterile nonabsorbent cotton
aqueous solution of saponin, 1 mg/ml (10 ml per laboratory section)
clinical centrifuge and centrifuge tubes
Pasteur pipets
chloroform (10 ml per laboratory section)
Sickle Cell Biokit (see Note)
Millipore Clinical Electrophoresis apparatus, including Power Module, PhoroSlide Single Cells (or Four Cells), PhoroSlide cellulose acetate strips, barbital buffer, sample applicators, accessory apparatus (1 cell per two students; see Note)
cytological staining dishes or Coplan jars (1 per two students)
TEB buffer: 16.1 grams Tris + 1.56 grams disodium EDTA + 0.92 gram boric acid in 1.0 liter distilled water, pH 8.65 (pH adjustment usually not necessary) (100 ml per two students)
parafilm squares
microscope slides and coverglasses
microscopes
n-propanol

Note: The Sickle Cell Biokit (74-4001) is available from Carolina Biological Supply Co., Burlington, NC 27215. One kit supplies sufficient materials for the solubility test and crisis simulation for a class of thirty students. No additional kit is required for the electrophoresis assays.

The Millipore apparatus is available from Millipore Biomedica, Acton, MA 01720. The Millipore technical pamphlet PM 701 gives detailed instructions in the use of the apparatus as well as many other applications of the technique.

Procedures

Phosphate Solubility Test
1. A simple solubility test can be used to differentiate between normal hemoglobin (hemoglobin A) and sickle cell hemoglobin (hemoglobin S). Hemoglobin S does not dissolve in a phosphate buffer solution and the suspension remains turbid, while hemoglobin A is soluble in the same solution and the mixture clarifies as the protein dissolves.
2. Pour 2.5 ml of the phosphate reagent solution supplied with the Carolina Sickle Cell Biokit into each of two plastic cups. To the first cup add 1 drop of normal blood; to the second cup add 1 drop of sickle cell trait blood. Mix each cup thoroughly with a clean toothpick, and place the cups over the lined circles on the data sheets provided with the kit. Periodically examine the contents of the cups.

Erythrocyte Morphology — Crisis Simulation
1. After you have cleansed your finger with 70% alcohol and allowed it to dry, pierce it with a sterile finger lancet and allow 2 drops of blood to fall onto a clean microscope slide, 1 drop at each end of the slide.
2. To one of the drops of blood, add a single drop of isotonic saline solution; to the second add a single drop of a sickling solution. Mix each drop with a clean toothpick. Place a clean coverglass over each of the drops and examine the slide under high magnification. Make accurate drawings of your erythrocytes under both conditions.
3. Examine the prepared microscope slide of sickle cell blood supplied with the Biokit. Draw representative erythrocytes.

Preparation of Hemoglobin Sample
1. Disinfect your finger and carefully puncture it with a sterile lancet. Collect 2 or 3 drops of blood in a clinical centrifuge tube. Immediately add 1 drop of the potent hemolytic agent saponin (1 mg/ml) and mix thoroughly. Next add 2 or 3 drops of chloroform and mix again. Centrifuge the tube in a benchtop clinical centrifuge at moderate speed for 5 minutes. After centrifugation, carefully remove the upper aqueous layer with a Pasteur pipet. This is a solution of your own hemoglobin, and it is ready for electrophoresis.
2. If sufficient quantities of sickle-cell-traited blood are available, you can also prepare a sample of this for electrophoresis. If only small quantities of this blood are available, the instructor will prepare it for you.

Electrophoresis
1. Carefully pipet 5.5 ml of the Millipore barbital buffer (pH 8.6) into each of the two chambers of the single cell. Be certain that the buffer level does not reach the top of the barrier between the chambers. Any buffer or moisture on the center barrier should be blotted dry. Note that this buffer can be used for two successive electrophoretic runs, but it should be discarded and replaced after the second experiment.
2. The PhoroSlides are packed between blue paper separators with the white cellulose acetate surface facing up. To remove the PhoroSlide from the box, use the MF forceps as shown in Figure 3-1A. Avoid touching the cellulose acetate surface with your fingers or with contaminating objects. The slides can be

FIGURE 3-1

A. Remove one PhoroSlide with the forceps supplied with the apparatus.

B. Flex the PhoroSlide with the metal holder and position it securely within the electrophoresis chamber. The buffer has already been added to the chamber.

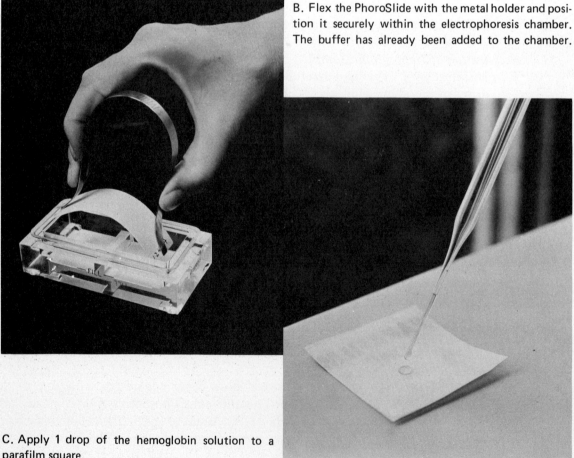

C. Apply 1 drop of the hemoglobin solution to a parafilm square.

Exercise 3] Electrophoresis of Human Hemoglobin 31

D. Load the sample applicator by passing it through the hemoglobin solution. Be certain that you do not allow the solution to fill the large space at the end of the applicator.

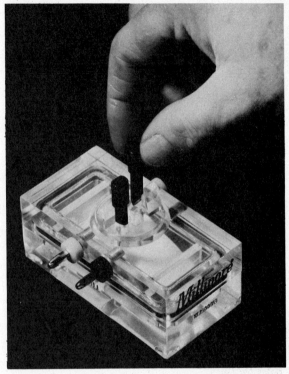

E. Add the applicator to the electrophoresis chamber. Allow it to touch the surface of the PhoroSlide but do not press it onto the surface.

F. Carefully attach the electrophoresis chamber to the power supply and carry out the procedure as directed.

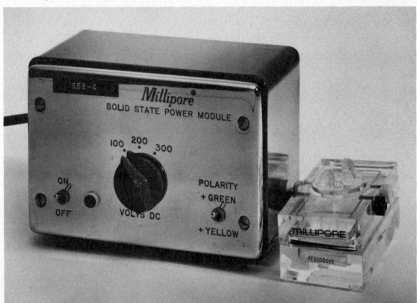

marked with the pen that is provided by holding the PhoroSlide under one of the blue separation papers.

3. Fill a clean cytological staining dish (or Coplan jar) with enough TEB buffer to cover the slides completely. Using the MF forceps, place several PhoroSlides in the staining dish. Be careful that adjacent slides do not contact each other. If tiny air bubbles form on the surface of the slides, try to remove them by gently tapping the staining jar on the bench top. Do not leave the PhoroSlides in the buffer for more than 15 minutes.

4. Use the MF forceps to remove the buffered PhoroSlides from the staining dish, and place each one on a clean blotter strip with the shiny Mylar surface down. Place a second clean blotter strip on the exposed cellulose acetate surface and press it firmly to absorb excess buffer. Allow the blotted PhoroSlide to dry in the air for 1 minute before placing it in the prepared electrophoresis cell.

5. With the forceps, lift the prebuffered PhoroSlide from the blotter pad. Flex the PhoroSlide as shown in Figure 3-1B and place it in the cell. When the slide is firmly seated, remove the forceps. Place the top portion of the electrophoresis cell firmly onto the bottom section. Place the applicator turntable in the circular recess on the cell cover with the arrow pointing to the "closed" position. Allow the cell and the PhoroSlide to equilibrate for 2 minutes.

6. Deposit a small sample of the hemoglobin solution onto a Parafilm square (Figure 3-1C). Do not leave it exposed for more than two or three minutes. Before drawing the sample into the tip of the applicator, be certain that the fine capillary is thoroughly clean and open. Draw the application tip through the drop of sample in the direction of the capillary space as shown in Figure 3-1D. Rotate the applicator turntable until the arrow is in the "open" position and carefully place the applicator into the slot (Figure 3-1E). Do not press the applicator onto the surface of the slide, but simply allow it to rest in position. A second sample can be applied. Add the sample of sickle cell trait blood to one side of the slide; add your own hemoglobin sample to the other side. Leave the applicators in contact with the PhoroSlides for two minutes to allow thorough transfer of the samples. Carefully remove the applicators and rotate the turntable to the "closed" position. The apparatus is now ready for electrophoresis.

7. Attach the electrophoresis cell to the Power Module (Figure 3-1F) and run the separation at a constant 100 volts for 27 minutes. Be certain to note carefully the electrical polarity of your electrophoresis run. Terminate the separation by turning off the power.

8. Remove the PhoroSlide strip from the cell following the separation.
   a. Place the slide in a solution tray containing Ponceau-S dye, at room temperature, for 10 minutes.
   b. Use the forceps to lift the PhoroSlide from the dye solution and allow the excess dye to drain off. Place it in the first of three solution trays containing 5% acetic acid for 2 minutes.
   c. Transfer the slide to the second acetic acid tray for an additional 2 minutes.
   d. Transfer the slide to the third acetic acid tray for an additional 2 minutes.
   e. Place the slide from the last rinse tray onto a clean blotter pad with the Mylar side down. Place a clean blotter on the exposed cellulose acetate surface and blot firmly. Remove the top blotter strip and allow the PhoroSlide to air-dry completely (the slide will turn completely white).
   f. Fill a staining dish with *n*-propanol or ethanol. Slowly immerse the dried PhoroSlide in the propanol and leave it for 5 minutes.

g. Transfer the PhoroSlide to clearing solution for 1 minute with the Mylar side down. Remove the slide from the clearing solution and allow it to dry thoroughly (preferably for several hours). *Do not attempt to blot your slide at this point! Disaster will result.*

9. After your stained slide has thoroughly dried, examine it and describe your results. How many hemoglobins are present? How far and in which direction(s) did they migrate? Summarize your findings on the data sheet at the end of this exercise.

STUDY QUESTIONS

1. What property of proteins makes it possible to separate them by electrophoresis? Which of the 20 amino acids are important in conferring this property on proteins?

2. What is the molecular basis for the difference in electrophoretic mobility between normal hemoglobin and hemoglobin S (the type found in sickle cell patients)? Explain your answer in detail.

3. Why did you carry out the electrophoretic separation at pH 8.65? Would the results have been different at pH 6? Explain.

4. Can all abnormal hemoglobins be diagnosed by electrophoresis?

5. What is the molecular basis by which it is thought cyanate may prevent the sickling of sickle cell hemoglobin?

REFERENCES

Allison, A. C. 1954. Protection afforded by sickle cell trait against subterian malarial infection. *Brit. Med. J., 1*:290.

Cerami, A., and C. M. Peterson. 1975. Cyanate and sickle cell disease. *Sci. Amer., 232*:44.

Culliton, B. J. 1972. Sickle cell anemia: The route from obscurity to prominence. *Science, 178*:138.

Green, R. L., R. G. Huntsman, and G. R. Serjeant. 1971. The sickle cell and altitude. *Brit. Med. J., 4*:593.

Ingram, V. M. 1958. Abnormal hemoglobins. I. The comparison of norman human and sickle cell hemoglobins by "fingerprinting." *Biochim. Biophys. Acta, 28*:539.

Levitan, M., and A. Montagu. 1971. *Textbook of Human Genetics.* Oxford University Press, New York and London.

Murphy, J. R. 1973. Sickle cell hemoglobin (Hb S) in black football players. *J. Am. Med. Assoc., 225*(Aug. 20, 1973):981.

Pauling, L., H. A. Itano, S. J. Singer, and I. C. Wells. 1949. Sickle cell anemia, a molecular disease. *Science, 110*:543.

Stamatoyannopoulos, G. 1972. Molecular basis of hemoglobin disease. *Ann. Rev. of Genetics, 6*:47.

EXERCISE 3

Name _____

Laboratory Section _____   Date _____

RESULTS AND CONCLUSIONS

Phosphate Solubility Test

Summarize your findings in this test with normal and sickle cell blood.

Erythrocyte Morphology—Crisis Simulation

Make accurate drawings of your erythrocytes incubated in isotonic saline solution and in the presence of sickling inducer. How does the chemical sickling agent induce sickling in normal erythrocytes?

Electrophoretic Separation of Hemoglobins

Summarize your results on the electrophoretic study of your own hemoglobin and sickle cell blood. Is hemoglobin A positively or negatively charged under the conditions of your assay? What is the net charge of hemoglobin S under these same experimental conditions?

SECTION II

The Heart

EXERCISE 4

The Cardiac Cycle in Man

OBJECTIVES In this exercise we shall familiarize ourselves with basic elements of the cardiac cycle, including pressure and volume changes in the heart, and use our understanding to interpret normal and abnormal sounds of the human heart.

The heart, as the central organ of circulation in the human body, functions continuously throughout life; its activity is, in many ways, synonymous with life itself. Two basic circulations exist. Pulmonary circulation delivers oxygen-deficient blood from the right side of the heart to the lungs where the blood is oxygenated and subsequently returned to the heart. General or systemic circulation carries blood from the left side of the heart through the great vessel, the aorta, to the organs and tissues of the body. Coordination of the two circulations, as well as rapid adjustment of heart activity to meet the ever-changing needs of the body, requires highly elaborate physiological controls.

The heart contains four cavities or chambers, two atria and two ventricles, as well as four valves (Figure 4–1). The mitral valve between the left atrium and left ventricle and the tricuspid valve between the right atrium and right ventricle are the two atrio-ventricular (A–V) valves. The aortic valve is situated between the left ventricle and the aorta; the pulmonic valve is found between the right ventricle and the pulmonary artery. All of the valves, with the exception of the mitral valve, have three cusps or leaflets; the mitral (bicuspid) valve has two. All four valves are continuous with the endocardium. The open ends of the two atrio-ventricular valves attach to delicate tendons, the chordae tendinae, and these in turn connect to the papillary muscles. The valves open only in one direction and this orientation determines the directional flow of the blood. Study Figure 4–1 and identify all of the structures mentioned above.

PRESSURE–VOLUME RELATIONSHIPS OF THE HEART

The detailed mechanics of heart action may be clarified by studying the pressure and volume relationships of the various chambers of the heart during the cycle. Cardiac physiologists distinguish between the left and right sides of the heart (or the left and right hearts, as they are sometimes termed). Similar pressure and volume relationships obtain for both sides, although the right heart is a region of lower blood

40 Section II] The Heart

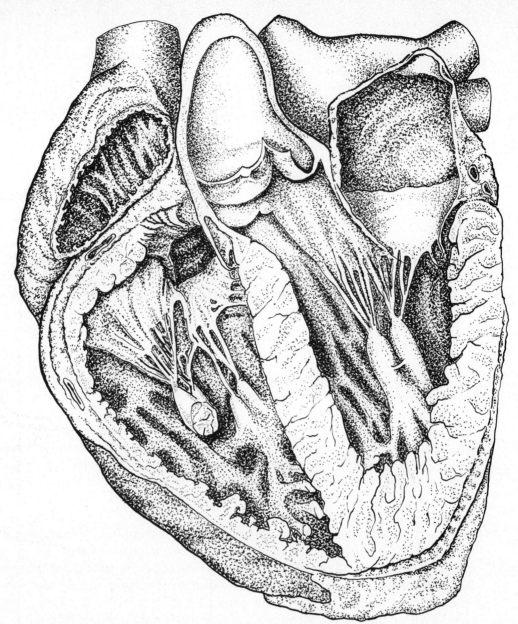

FIGURE 4-1 The human heart. Identify and label the various chambers, valves, and major blood vessels.

pressure and much lower oxygen tension. In the following discussion we shall consider only the left heart, but our analysis can also be applied to understanding of right heart function.

Figure 4-2 depicts several of the processes that occur during a typical cardiac cycle. Changes in atrial pressure, ventricular pressure, ventricular volume, and aortic pressure are presented. Study this diagram carefully and use it as a basis for the discussion that follows.

FIGURE 4-2 [opposite] Relationship of pressure, volume, and sounds of the heart during the cardiac cycle. The time scale is represented along the abscissa.

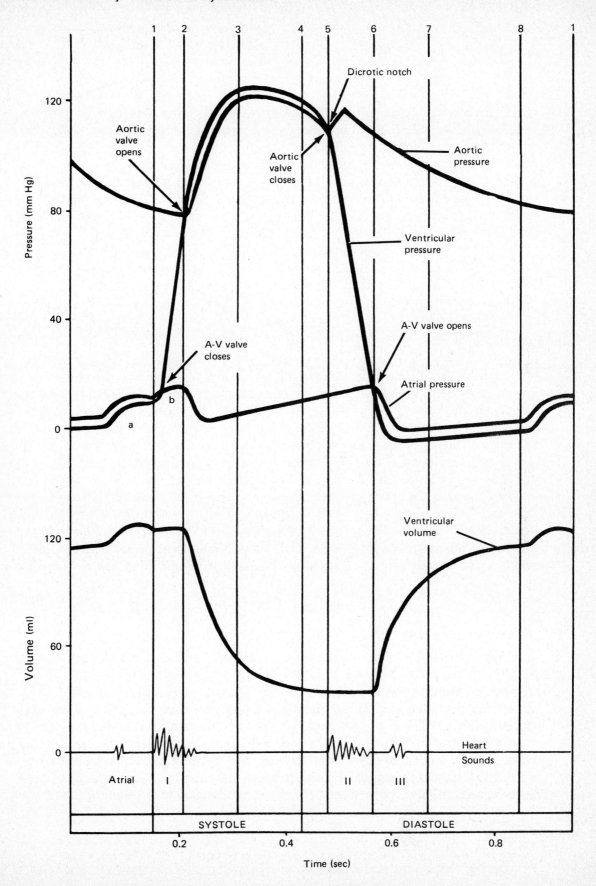

Atrial Pressure

Blood pressure within the left atrium is quite low and varies from 0 to about 15 mm Hg over the course of the heart cycle. During the relaxed portion of the cycle (*diastole*), pressure within the left atrium steadily increases as blood enters the chamber from the pulmonary system. Shortly before contraction occurs, there is a small, abrupt increase in atrial pressure caused by atrial contraction, or *atrial systole*. (The word *systole,* unless modified, ordinarily refers to ventricular contraction.) At this time the mitral valve is open, the aortic valve is closed, and the ventricles are largely filled with blood. After systole begins (vertical line 1 in Figure 4-2), the pressure within the ventricles increases rapidly and this enhanced pressure closes the mitral valve. Following mitral valve closure, the atrial pressure gradually increases as the atria fill with blood (lines 2-6).

When the intraventricular pressure has fallen sufficiently, the increased pressure within the left atrium opens the mitral valve (line 6), blood flows through the atrium into the left ventricle, and the atrial pressure decreases. Filling of the ventricle occurs and the cycle is complete.

Ventricular Pressure

At the onset of ventricular systole, the aortic valve is closed. Closure of the mitral valve makes the ventricle a closed chamber and the intraventricular pressure increases rapidly (from about 10 to 80 mm Hg) as the ventricle contracts. Increased ventricular pressure causes the aortic valve to open; blood flows from the heart into the aorta and arterial system in the rapid ejection phase (between lines 2 and 3 in Figure 4-2). At line 4 much of the blood has left the ventricle and the intraventricular pressure begins to fall. The aortic valve closes at line 5, and once again the ventricle is a closed chamber. In the meantime the atria have been filling and the atrial pressure is gradually increasing. At line 6 the mitral valve opens and blood flows into the ventricles once again. "Rapid inflow" filling of the ventricles takes place (between lines 6 and 7) and continues until atrial systole (line 8). The relatively quiescent period between lines 7 and 8 is usually called diastasis, which means "a separation," in this case, of cardiac cycles.

Ventricular Volume

Changes in ventricular volume are also illustrated in Figure 4-2. During diastole the aortic valve is closed, the mitral valve is open. Blood, returning to the heart, passes into the left ventricle and this chamber fills to a volume of 125 ml. Ventricular systole closes the mitral valve and, for a brief interval, both valves are closed (between lines 1 and 2). During this time the ventricular volume must remain constant because there is no blood leaving the chamber. As the pressure within the ventricle increases, the aortic valve opens and blood flows from the heart into the aorta. About 80 ml flows from the ventricle, leaving a residual volume of 40-45 ml. After the aortic valve closes (line 5) and the mitral valve opens (line 6), the ventricle fills again to a volume of 125 ml.

Aortic Pressure

During diastole the aortic pressure gradually decreases, as shown in Figure 4-2, because the aortic valve is closed and no blood is entering the aorta and arterial system. The aortic pressure decreases steadily to about 80 mm Hg, at which time the intraventricular pressure, which is rapidly increasing in systole, opens the aortic valve (line 2). The aortic pressure rapidly increases to the systolic level (approximately 120 mm Hg). The transient increase in aortic pressure, known as the dicrotic notch (between lines 5 and 6), is caused, in part, by elastic recoil of the aorta following closure of the aortic valve.

HEART SOUNDS

The beating of the heart produces two distinct sounds that can be heard with a stethoscope. Two other inaudible sounds can be discerned and recorded with suitable electronic equipment. The cardiac sounds are often expressed by the monosyllables "lub-dup," and the cycle is characterized as lub-dup, pause, lub-dup, pause. Lub-dup is the systolic portion of the cycle, and pause represents diastole.

The first heart sound, lub, is caused by several factors: (1) closure of the atrioventricular valves and tensions set up in the valvular leaflets and chordae tendinae as the intraventricular pressure rises, (2) contraction of the ventricles themselves, and (3) oscillations of the heart and large vessels due to changes in blood flow resulting from heart activity. Vibrations of the ventricles and other elements of the first sound are transmitted through surrounding tissue to the chest wall where they can be heard and recorded. The first sound is low-pitched, dull, and fairly long: it is primarily a function of the force of ventricular contraction. Note in Figure 4-2 the time at which this sound occurs.

The second heart sound, dup, defines the end of systole and is generally associated with closure of the aortic and pulmonic valves. The second sound is ordinarily clear, sharp, and higher in pitch. During inspiration there may be a short interval between closure of the aortic and pulmonic valves so that the second sound is "split."

The third sound is usually not heard with the unaided ear. It occurs during rapid-inflow filling of the ventricles and is thought to reflect vibrations of the ventricular walls as blood rushes into the chambers. Examine Figure 4-2 to note the time at which this sound occurs.

The atrial sound is also inaudible and is generally obscured by the first heart sound. It reflects both atrial systole and the flow of blood from the atria into the ventricles.

In this exercise we shall listen to our own heart sounds with a stethoscope, and study supplemental materials dealing with normal and abnormal heart sounds.

Materials

stethoscopes (1 per two to four students)
materials for studying abnormal heart sounds (see Note)

Note: A series of 33⅓ rpm records entitled *Interpreting Heart Sounds* is available from Roche Laboratories, Division of Hoffman-LaRoche, Nutley, NJ 07110. These are also available on free loan from local chapters of the American Heart Association. The recordings can be studied in a language laboratory or other suitable facility. An excellent series of audiotapes dealing with heart sounds and murmurs (WG141 H436) is available from the Library, College of Medicine and Dentistry of New Jersey, 100 Bergman St., Newark, NJ 07103. Finally, the Syracuse University Film Library lists an appropriate 16 mm film entitled *Heart Disease: Its Major Causes* (see page 337).

Procedures

1. Remove the outer clothing over the region of the heart. Female students may listen to the sounds of the heart in a suitably private area. You may listen to your own heart or use your laboratory partner as a subject.
2. With the stethoscope locate the apex beat in the fifth intercostal space on the left side, as shown in Figure 4-3. Place the bell of the stethoscope lightly but firmly over this area and listen to the characteristic sounds. Next, place the bell in the second intercostal space on both the right and left sides of the sternum. The aortic and pulmonic valves can be heard at these positions.
3. Have the subject exercise vigorously for a short time and repeat the observations. Note also the effects of breathing on the heart sounds.
4. After you have thoroughly acquainted yourself with the heart sounds, study the supplemental materials that have been provided dealing with abnormal heart sounds.

FIGURE 4-3 Positions for proper application of the stethoscope to auscultate the sounds of the human heart.

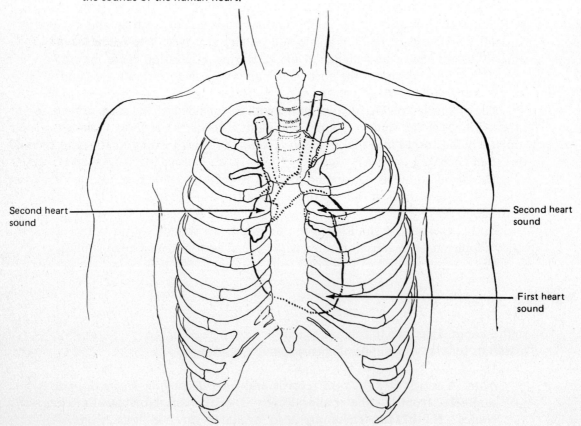

Exercise 4] The Cardiac Cycle in Man

STUDY QUESTIONS

1. Examine Figure 4–2 and determine what fraction of the heart cycle is systolic, what fraction is diastolic. What constitutes the beginning of systole? The endpoint of systole? (The heart rate of the individual depicted in Figure 4–2 is 72/minute; the length of the cycle is 60 seconds/72 beats = 0.83 seconds/beat).

2. The mitral and aortic valves are both closed during a certain portion of the cardiac cycle. Are these valves both open at any point in the cycle?

3. Although the overall features of left- and right-heart circulation are generally similar, differences do exist in detail. List as many differences as you can between the two circulations. How are these differences reflected in the anatomical structure of the heart?

4. What are the specific functions of the papillary muscles and the chordae tendinae in the operation of the heart valves? Are there defects known in which these structures fail to operate properly? What consequences does this have?

5. What differences exist in the functioning of the heart during fetal life? What are the circulatory changes that occur after birth and how long is it before the heart takes on its usual structure and function? Characterize at least one heart defect in which the switchover does not occur normally.

6. How does the heart itself receive a supply of blood? What happens if this supply is deficient? What is this condition called and what are its causes?

7. Assume that the stroke volume of the heart for a resting individual is 80 ml. What is the cardiac output (minute volume) for this individual? What percentage of the total blood volume in the body does this represent? What is the Fick method for determining cardiac output and how is the procedure performed? Can you provide a sample calculation showing typical numerical values that one might actually obtain?

8. What is Starling's Law of the Heart? What significance does it have for the functioning of the heart?

9. Some common heart defects are

Valve	Abnormality
Aortic or pulmonic	Stenosis
Aortic or pulmonic	Insufficiency
Mitral or triscuspid	Stenosis
Mitral or tricuspid	Insufficiency

Characterize each of the defects listed above and describe the alteration that would occur in the heart sounds.

REFERENCES

Armour, J. A., and W. C. Randall. 1970. Structural basis for cardiac function. *Am. J. Physiol., 218*:1517.

Ebert, J. P. 1959. The first heartbeats. *Sci. Amer., 200*(3):87.

Guyton, A. C. 1976. *Textbook of Medical Physiology,* 5th ed. W. B. Saunders, Philadelphia.

Louisada, A. A. 1972. *Sounds of the Normal Heart.* Green, St. Louis.

Louisada, A. A., et al. 1974. Changing views on the mechanism of the first and second heart sounds. *Am. Heart J., 88*:503.

McKusick, V. A. 1956. Heart sounds. *Sci. Amer., 194*(5):120.

Salk, L. 1973. The role of the heartbeat in relations between mother and infant. *Sci. Amer., 228*(5):24.

Willius, F. A., and T. E. Keys (eds.). 1961. *Classics of Cardiology* (2 vols.). Dover, New York.

EXERCISE 4

Name _____

Laboratory Section _____   Date _____

RESULTS AND CONCLUSIONS

Normal Heart Sounds

Summarize your observations on basic heart sounds. What differences did you note at the different chest positions? What effects (if any) did you notice of breathing on the heart sounds? What effects did exercise have? Were both sounds affected equally?

Abnormal Heart Sounds

Summarize your observations and findings on abnormal heart sounds. How are these used in cardiac diagnosis?

EXERCISE 5

Human Blood Pressure

OBJECTIVES In this exercise we shall learn to use the sphygmomanometer in measuring arterial blood pressure. Various factors influencing blood pressure, including postural changes, cold, and vigorous exercise, will be studied. Comparisons will be made between well-conditioned and poorly conditioned students, between smokers and nonsmokers.

With each beat of the heart, a volume of blood flows from the left ventricle into the aorta and arterial system. The rhythmic flow causes the blood pressure to rise and fall during each beat. The highest pressure during the cycle is termed the *systolic* pressure; the lowest value is *diastolic* pressure. Blood pressure is defined as that pressure exerted against any unit area of the surface of a blood vessel. It is ordinarily expressed as millimeters of mercury (mm Hg), because the mercury manometer has long been used in clinical testing. A reading of 150 mm Hg at a given point in the cardiovascular system means that the blood exerts sufficient pressure at this point to raise a column of mercury to a height of 150 mm.

Many factors operate to regulate blood pressure within the human body. Arterial pressure can be shown to be directly proportional to the output of heart (i.e., volume of blood per unit time) and the peripheral resistance.

$$\text{Pressure} \propto \text{Cardiac output} \times \text{Peripheral resistance}$$

Obviously, an increase in peripheral resistance (such as the constriction of many small blood vessels) for a given cardiac output causes an immediate increase in blood pressure. Since more than half of the total flow resistance in the human cardiovascular system occurs in the arterioles, these vessels are extremely important in regulating blood pressure. Factors controlling cardiac output (heart rate and stroke volume) are also significant.

Hypertension, or high blood pressure, is a significant health problem and is related to abnormalities of arterial pressure regulation. High blood pressure and its consequences are among the leading causes of death in the United States. The condition may accelerate the development of arteriosclerosis and can eventually result in congestive heart failure and other forms of heart disease, stroke, and kidney disease. Unfortunately, hypertension produces no obvious external symptoms and may remain unrecognized for many years. Thorough discussions of the diagnosis and management of hypertension are found in several of the references listed at the end of this exercise.

TABLE 5-1 Blood Pressure Values for Men and Women[a]

	Systolic			Diastolic		
Age	Normal Range	Mean	Hypertension Lower Limit	Normal Range	Mean	Hypertension Lower Limit
			MEN			
16	105-135	118	145	60-86	73	90
17	105-135	121	145	60-86	74	90
18	105-135	120	145	60-86	74	90
19	105-140	122	150	60-88	75	95
20-24	105-140	123	150	62-88	76	95
25-29	108-140	125	150	65-90	78	96
30-34	110-145	126	155	68-92	79	98
35-39	110-145	127	160	68-92	80	100
40-44	110-150	129	165	70-94	81	100
50-54	115-160	135	175	70-98	83	106
55-59	115-165	138	180	70-98	84	108
60-64	115-170	142	190	70-100	85	110
			WOMEN			
16	100-130	116	140	60-85	72	90
17	100-130	116	140	60-85	72	90
18	100-130	116	140	60-85	72	90
19	100-130	115	140	60-85	71	90
20-24	100-130	116	140	60-85	72	90
25-29	102-130	117	140	60-86	74	92
30-34	102-135	120	145	60-88	75	95
35-39	105-140	124	150	65-90	78	98
40-44	105-150	127	165	65-92	80	100
45-49	105-151	131	175	65-96	82	105
50-54	110-165	137	180	70-100	84	108
55-59	110-170	139	185	70-100	84	108
60-64	115-175	144	190	70-100	85	110

[a]Reprinted with permission from Eli Lilly and Company, Indianapolis, Indiana.

Table 5-1 lists blood pressure values for men and women as a function of age.

SPHYGMOMANOMETRY

The sphygmomanometer is a familiar instrument used in the clinical evaluation of blood pressure. An inflatable cuff is used to occlude circulation, usually in the brachial artery of the arm. The cuff pressure is then released very slowly and the examiner listens with a stethoscope for characteristic sounds as circulation is resumed in the artery. Figure 5-1 illustrates the correct position of the cuff and shows the underlying brachial artery.

Exercise 5] *Human Blood Pressure*

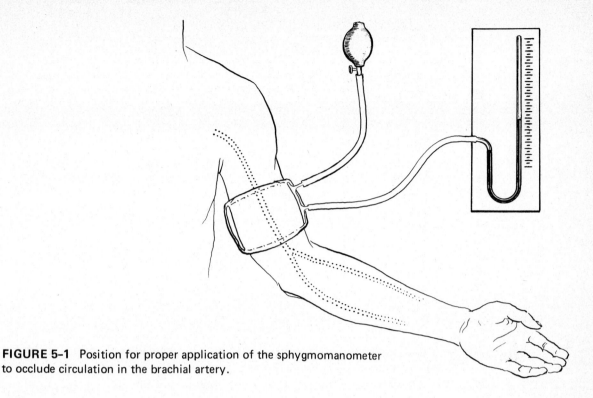

FIGURE 5-1 Position for proper application of the sphygmomanometer to occlude circulation in the brachial artery.

Materials

    sphygmomanometers (1 per two to four students)
    stethoscopes (1 per two to four students)
    folding cots (or flat laboratory benches)

Procedures

1. Wrap the cuff around the subjects's arm above the elbow with the small inflatable bag on the median surface. The end flap of the cuff should be tucked under the previous fold.
2. Feel the radial pulse by hand as you inflate the cuff to 160-170 mm Hg (i.e., above the subject's anticipated systolic pressure). Release the pressure in the cuff very slowly and palpate the radial artery. The point at which the pulse is restored is the *systolic pressure*. Repeat the procedure to validate your determination.
3. Next, place the bell of the stethoscope over the brachial artery above the elbow joint (see Figure 5-1) and secure it under the cuff. Inflate the cuff beyond the systolic pressure as before. Slowly open the needle valve and listen carefully for the Korotkow sounds. The first pulse wave that passes through the brachial artery produces a distinct sound (very often a sharp thudding); the pressure at which this occurs is the *systolic pressure*.
4. Slowly release the pressure below the systolic point and listen very carefully. The sounds increase and reach maximal intensity as the pressure drops over a range of 30-40 mm Hg, then become muffled and finally disappear. The point at which

the sound becomes muffled is usually several mm higher than the final point of disappearance and is often taken as the *diastolic pressure,* although the American Heart Association defines diastole as the point of disappearance. Blood pressure is recorded as systole/diastole, e.g., 120/80; or as systole/diastole/end diastole, e.g., 120/80/75.

5. Repeat the procedure until you can obtain reliable values. The cuff should be inflated around the subject's arm for only short periods of time. Checks can be made by testing both arms.

EFFECT OF POSTURE ON BLOOD PRESSURE

Materials

Same as for sphygmomanometry

Procedures

1. Take the blood pressure of the subject after he has reclined for 5 minutes. Make several determinations and average the results.
2. Repeat the measurements immediately after the subject has assumed a standing position. Repeat after 2, 5, and 10 minutes. Record the data.

COLD PRESSOR TEST

Cold stimuli have a well-known vasoconstricting effect. The response in man to extreme cold is an increase in systolic blood pressure of about 10 mm Hg. An individual with a tendency toward hypertension will show a much more pronounced increase, occasionally approaching 40 mm. A value of 23 mm or higher is often regarded as a possible indication of hypertension.

Materials

sphygmomanometers (1 per two to four students)
stethoscopes (1 per two to four students)
stop watches (or wristwatches)
containers of water at 5° C (1 per four to eight students)
folding cots (or flat laboratory benches)

Procedures

1. Measure the blood pressure of the subject in a sitting position. Take several measurements to insure accuracy.
2. Have the subject immerse one hand in 5°C water and after a short time measure the pressure again. Follow it over a period of 2 minutes. What do you note?

Exercise 5] Human Blood Pressure

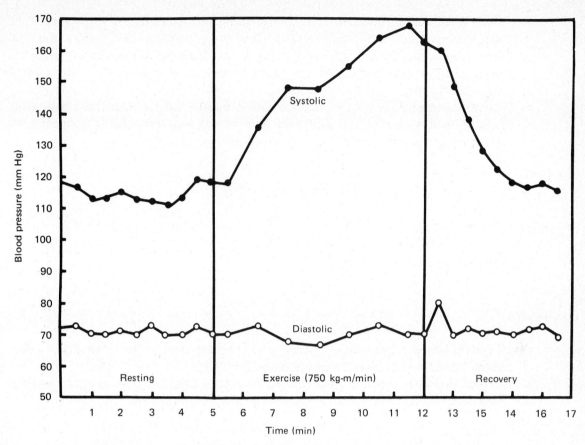

FIGURE 5-2 A typical time course for the arterial blood pressure response to rest, exercise, and recovery in a healthy young man.

EFFECT OF EXERCISE ON BLOOD PRESSURE

During rhythmic (isotonic) exercise involving fairly heavy work loads, there is usually a dramatic increase in systolic pressure, but little change in diastolic pressure. (What changes would you expect in cardiac output and peripheral resistance during exercise?) Figure 5-2 illustrates the effect of a 7 minute ride on the bicycle ergometer at a work load of 750 kg-m/minute. **Students with high blood pressure should not participate in this portion of the exercise.**

Materials

sphygmomanometer
stethoscope
bicycle ergometer (or stepping benches)
stopwatches (or wristwatches)

Procedures

1. Determine the subject's blood pressure with the subject standing quietly (if he is to perform bench stepping) or seated quietly on the ergometer (if he is to pedal).

2. Pedal for 3 minutes at a work load of 600-750 kg-m/minute, or bench step for 2 or 3 minutes at 30 steps/minute. Directly after the exercise period, determine the subject's blood pressure. Repeat these measurements following exercise at 1 minute intervals until control valves are recorded. How long did the recovery period take? Did you notice any obvious physiological signs in the subject indicating elevated blood pressure? What? This test can be performed with well-conditioned and poorly conditioned individuals and the results compared. A comparison of smokers and nonsmokers is also of interest.

STUDY QUESTIONS

1. What effect does hardening of the arteries have on blood pressure? Why should retention of salt in the blood have an effect on blood pressure?

2. What medications are used to control hypertension? How do they act?

3. What is renal hypertension and how does it cause elevated blood pressure? How can this condition be treated?

4. What is postural hypotension and what is its underlying cause?

5. Explain the role of the autonomic nervous system in regulating blood pressure.

REFERENCES

Barcroft, H., and H. J. C. Swan. 1953. *Sympathetic Control of Human Blood Vessels*. E. Arnold and Co., London.
Burton, A. C. 1953. Peripheral circulation. *Ann. Rev. Physiol., 15*:213.
Franz, G. N. 1974. On blood pressure control. *Physiologist, 17*:73.
Geddes, L. A. 1970. *The Direct and Indirect Measurement of Blood Pressure*. Yearbook Medical Publishers, Chicago.
Guyton, A. C., et al. 1972. Circulation: Overall regulation. *Ann. Rev. Physiol., 34*:13.
Kotchen, J. M. 1974. Blood pressure distributions in urban adolescents. *Am. J. Epidemiology, 99*:315.
Laragh, J. H. (ed.). 1973. Symposium on hypertension: Mechanisms and management. *Am. J. Med., 55*:261.
Page, I. H. 1948. High blood pressure. *Sci. Amer., 179*(4):44.
Steinfeld, L., et al. 1974. Updating sphygmomanometry. *Am. J. Cardiol., 33*:107.
Zweifach, B. W. 1959. The microcirculation of the blood. *Sci. Amer., 200*(1):54.

EXERCISE 5

Name _____

Laboratory Section _____ Date _____

RESULTS AND CONCLUSIONS

Indirect Blood Pressure Determination (Sphygmomanometry)

    Systolic pressure (radial palpation)_____

    Systolic/diastolic (auscultation)_____

    Examine the data in Table 5-1 and note how your blood pressure compares with individuals of your own age and sex.

Effects of Posture

    Reclining pressure_____

    Standing (immediate)_____

    Standing (2 minutes)_____

    Standing (5 minutes)_____

    Standing (10 minutes)_____

    How can you explain these findings?

Cold Pressor Test

    Control pressure (sitting)_____

    Following cold stimulus_____

What conclusions can you draw from this test?

Effects of Exercise

Summarize your results dealing with the effects of exercise on systolic and diastolic pressures. Collect data from other members of the class and interpret them. What are the differences between well-conditioned and poorly conditioned students? Between smokers and nonsmokers?

EXERCISE 6

Electrocardiography

OBJECTIVES In this exercise we shall familiarize ourselves with elementary aspects of the theory and practice of electrocardiography. Measurements will include (1) standard EKG with the subject in a supine position and (2) EKG under various experimental conditions, e.g., after moderate exercise, after smoking a cigarette or drinking coffee, accompanying certain postural changes, etc. Thorough analysis of the electrocardiogram will be conducted.

The functioning of the human heart is accompanied by weak electrical activity, initiated at the sino-atrial (S-A) node, and subsequently transmitted throughout the heart. Arrival of the electrical signal at the contractile fibers of the heart initiates contraction. The currents eventually spread through the entire human body and may be recorded with suitable detecting apparatus. Electrocardiography, the measurement and analysis of currents associated with heart activity, is extremely important in the diagnosis of abnormal cardiac function.

ELECTRICAL ACTIVITY OF THE HEART

The dominant pacemaker of the heart, the sino-atrial (S-A) node, lies in a cellular cluster in the right atrial wall between the inlets of the inferior and superior vena cavae (Figure 6-1). Impulses originating from this node spread to adjacent myocardial cells and are propagated as a wave of depolarization through preferential atrial pathways (not shown in Figure 6-1). Contraction of the muscular walls of the atria immediately follows atrial depolarization.

The atrio-ventricular (A-V) node, located within the lower intra-atrial septum, relays the impulse from the atria to the ventricles. After the impulse passes through the A-V node, it enters the fiber tract known as the common bundle or the bundle of His. This short pathway subdivides into right and left bundle branches. The smallest conductive elements, the Purkinje fibers, are distributed throughout the inner walls of the ventricles, throughout the muscular septum separating the two ventricles, and within the muscular papillae where they make intimate contact with contractile elements of the heart. After being disbursed throughout the entire inner surface of the ventricles, the cardiac impulse moves toward the outer surface of the heart as a generalized wave of depolarization.

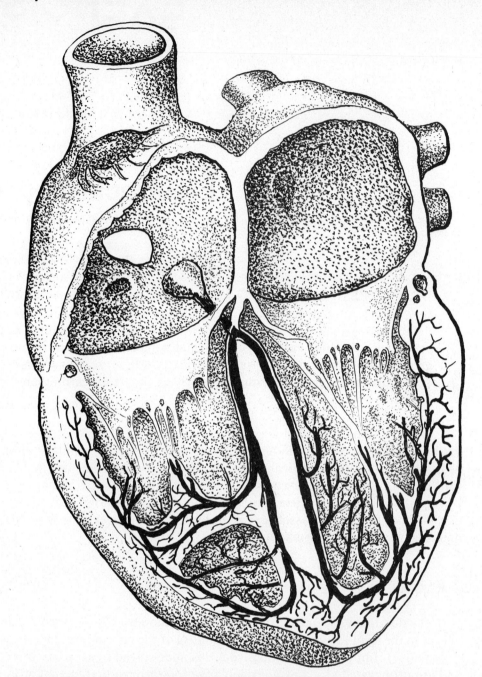

FIGURE 6-1 Drawing of the human heart illustrating neural pathways of the heart. Identify and label the various components.

BASIC ELECTROCARDIOGRAPHY

Electrocardiography assumes the human body to be a volume conductor consisting of electrolytes in aqueous solution. The heart is seen as a dipole, i.e., a pair of point sources of electrical charge equal in magnitude but opposite in sign. Current flows between the two poles throughout the volume conductor. Detecting electrodes placed anywhere on the body, therefore, are influenced by these currents.

Various electrode positions are used to produce an electrocardiogram. Einthoven selected three standard positions: the left forearm (LA); the right forearm (RA); and the left lower leg (LL). A fourth electrode is sometimes attached to the right leg (RL), but this serves only as a ground connection. Additional EKG recordings are made with a fifth chest electrode which is placed at various positions on the chest.

The detecting electrodes are usually constructed of metal and are slightly concave to insure good contact with the skin. EKG solution, a jelly containing an electrolyte such as KCl or NaCl, is rubbed over the skin. This abrasive action removes dead skin cells and other interfering materials and the electrolyte solution improves the electrical conductivity of the skin which has a fairly high resistance. The electrodes are held in place with snugly fitting rubber straps.

The particular arrangement of two electrodes is known as a lead and, by convention, three standard leads have been established (Figure 6-2).

Lead I. The negative terminal of the EKG machine is connected to the electrode on the right arm, the positive terminal to the left arm. Accordingly, when the right arm is negative with respect to the left arm, there is a positive or upward deflection of the recording pen.

Lead II. The negative terminal is connected to the right arm, the positive terminal to the left leg.

Lead III. The negative terminal is connected to the left arm, the positive terminal to the left leg.

The relation between the three leads is expressed algebraically as Einthoven's equation: Lead II = Lead I + Lead III.

FIGURE 6-2 The Einthoven triangle superimposed on a human subject. Three of the standard electrocardiographic leads are illustrated.

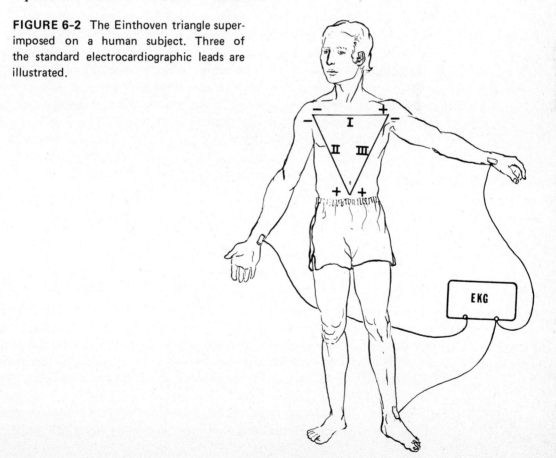

The cardiac impulse is initiated at the S-A node, a region of the heart that is closer to the RA electrode than to the LA electrode. The impulse may be considered as a wave of negativity, the right arm thus becomes negative prior to the left arm, and the deflection is positive. (Refer back to the definition of lead I above, if this is unclear.) A positive response on lead I is recorded, by convention, as an *upward* deflection of the pen. The P wave, reflecting depolarization of the atria, is a good example of this process. (Can you show why the P wave is also positive in lead III EKG readings?)

THE ELECTROCARDIOGRAM

Figure 6-3 illustrates the components of a normal electrocardiogram as measured on lead II.

(1) *P wave:* The upward (positive) deflection associated with *atrial depolarization*. The average duration is 0.08 second (range 0.06-0.12 second), and the amplitude is not greater than 0.3 millivolt.

(2) *QRS complex:* The series of negative and positive deflections associated with *ventricular depolarization*. It is measured as the portion of the tracing from the beginning of the Q wave to the end of the S wave. Repolarization of the atria takes place simultaneously and is obscured by the ventricular changes. As measured on lead II, the average duration is 0.08 second (range 0.06-0.10 second) and the amplitude of the R wave is not over 2.5 millivolts.

(3) *PR interval:* The portion of the tracing from the initiation of the P wave to the beginning of the QRS complex. In this interval an impulse passes from the S-A node through the atrial muscle, the A-V node, and the A-V bundles. The average duration is 0.13-0.16 second. Any lengthening of the PR interval beyond about 0.20 second indicates an impairment of impulse conduction. In complete heart block the initial P wave may not be followed at all by the QRS complex (i.e., no ventricular contraction occurs).

(4) *T wave:* The positive deflection (on leads I and II) following the QRS complex representing ventricular repolarization. The T wave is of longer duration and lower amplitude than the initial depolarization (QRS complex), which suggests that the ventricular repolarization is slower and less well synchronized. If the T wave is inverted (negative) on leads I and II, this finding is considered to be abnormal. An inverted T wave on lead III is fairly common and is not abnormal. The average duration of the T wave is about 0.16 second; the amplitude is about 0.3 millivolt but is highly variable.

(5) *PR segment:* The portion of the tracing from the end of the P wave to the beginning of the QRS segment. Lengthening of this segment beyond about 0.12 second may indicate malfunction of the A-V node or the bundles of His.

(6) *ST segment:* The portion of the tracing between the end of the S wave and the beginning of the T wave. This segment ordinarily lasts about 0.08 second and is usually level with the PR segment. Upward or downward displacement may indicate damage to the cardiac muscle or strain on the ventricles. A good discussion of this aspect of the EKG is given in Must [1968].

(7) *QT interval:* The interval from the beginning of the QRS complex to the end of the T wave. This interval is usually referred to as electrical systole. At a

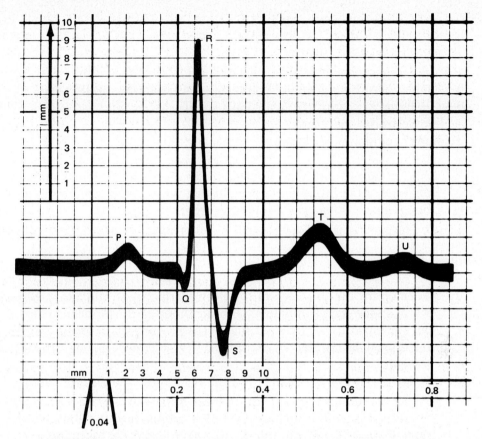

FIGURE 6-3. A tracing on lead II of the normal EKG cycle. The recording speed is 25 mm/second. One large division along the X-axis represents 0.20 second. The vertical excursion of the pen is 1 millivolt/cm.

heart rate of 70 beats per minute the QT interval is about 0.36 second; at higher heart rates the interval is shorter.

Electrocardiographic analysis reveals a great deal about the detailed functioning of the heart in both normal and abnormal states. If, for example, the left fiber bundle is diseased so that conduction along this route cannot take place, left ventricular contraction is greatly delayed (and occurs only after a detoured wave of depolarization arrives from the right ventricle). Such a condition of left bundle block may be diagnosed electrocardiographically. In other forms of heart diseases the cardiac impulse is blocked at the A-V node. Under these circumstances the ventricles beat at their own inherent rate (slower than atrial contraction), and such a condition of heart block is evident in the EKG recording. For a detailed discussion of many other heart defects and the use of electrocardiography in cardiac diagnosis see Goldman [1973] and Phillips and Feeney [1973].

Stress testing is often included in the clinical analysis of heart function. An individual who shows an otherwise normal EKG pattern may demonstrate significant abnormalities when stressed through vigorous exercise to 90% of his maximum heart rate. Because of the potential dangers to some individuals inherent in stress testing, evaluations of this kind are ordinarily performed only in the presence of a physician. Ellestad [1975] and Wilson [1975] provide lucid discussions of graded stress tests.

Materials

electrocardiograph (see Note)
electrocardiograph paper
lead cables: RA (right arm), RL (right leg), LA (left arm), LL (left leg).
EKG solution or paste (or 70% alcohol)
4 plate electrodes and straps
folding cot or laboratory bench

Note: Many commercial models are available. In order to make detailed analyses of the EKG, it is important that the electrocardiographic apparatus be equipped with a suitable chart recorder.

Procedures

1. Turn on the power switch of the electrocardiograph and allow the instrument to warm up for several minutes. Calibrate the apparatus so that a pen deflection of 1 cm corresponds to 1 millivolt (mV). The procedures for doing this depend on the instrument being used; more detailed instructions will be provided by the instructor.
2. Select the sites for attachment of the electrodes, and cleanse the skin thoroughly with EKG solution or 70% alcohol. The palmar aspects of the wrists and the inner surfaces of the ankles are the appropriate locations. Place a small thickness of gauze pads wetted with EKG solution or alcohol between the skin and the plate electrode. Secure the electrodes in a snug position with the rubber straps.
3. Connect the lead cables to the electrodes and to the EKG apparatus. The lead adaptor (or the switch on the EKG apparatus) should be in the standby position until the actual EKG recordings are made.
4. Measure a standard EKG with the subject in the supine position. Measure Lead I (LA-RA), Lead II (LL-RA), and Lead III (LL-RA). The chart speed should be set at 25 mm/second.
5. EKG measurements can also be made under a variety of other conditions. Perform as many of the following as time allows:
   a. (sitting position) EKG during deep prolonged inspiration; the breath is held for approximately 30 seconds. Be certain to notice the heart rate before, during, and after breath holding. This can be done as a separate determination using a slower chart speed on the recorder.
   b. EKG during the Valsalva maneuver (forced expiration with closed mouth, glottis, and nose to make ears "pop").
   c. EKG immediately on standing and after 1 minute of standing.
   d. EKG after smoking a cigarette or drinking strong coffee.
   e. EKG immediately after moderate exercise. Perform 2 minutes of bench stepping at a rate of 30 steps/minute. (Use a 20 inch bench for male students, an 18 inch bench for females.) Measure the post-exercise EKG with the subject in a supine or sitting position.

STUDY QUESTIONS

1. Characterize the following cardiac conditions and show how each may be diagnosed in the EKG: complete and partial heart block; tachycardia; bradycardia; atrial arrhythmia; ventricular arrhythmia. How is the EKG used to diagnose and localize areas of myocardial infarction?

2. What are some of the abnormalities that may appear in the EKG after stress testing?

3. Draw a diagram similar to Figure 4-2 in the previous exercise in which you include the EKG responses in addition to the heart sounds and other parameters included in the diagram. Be certain to indicate accurately the precise times during the cardiac cycle when the electrocardiographic changes can be detected.

REFERENCES

Burch, G. E., and N. P. DePasquale. 1964. *A History of Electrocardiography.* Year Book Med. Pub., Chicago.

Ellestad, M. H. 1975. *Stress Testing: Principles and Practice.* F. A. Davis, Co., Philadelphia.

Goldman, M. J. 1973. *Principles of Clinical Electrocardiography,* 8th ed. Lange Med. Pub., Los Altos, CA.

Hantzsche, K., and K. Dohru. 1966. The electrocardiogram before and after a marathon race. *J. Sports Med., 6*:28.

Kozar, A. J., and P. Hunsicker. 1963. A study of telemetered heart rate during sports participation of young adult men. *J. Sports Med., 3*:1.

Langer, G. A. 1973. Heart: Excitation-contraction coupling. *Ann. Rev. Physiol., 35*:55.

Master, A. M., et al. 1942. The electrocardiogram after standard exercise as a functional test of the heart. *Am. Heart J., 24*:777.

Must, A., et al. 1968. Exercise ST changes in healthy young men. *Arch. Internal Med., 121*:22.

Phillips, R. E., and M. K. Feeney. 1973. *The Cardiac Rhythms, A Systematic Approach to Interpretation.* W. B. Saunders, Philadelphia.

Scher, A. M. 1961. Electrocardiogram. *Sci. Amer., 205* (5):132.

Weidmann, S. 1974. Heart: Electrophysiology. *Ann. Rev. Physiol., 36*:155.

Wilson, P. K. (ed.). 1975. *Adult Fitness and Cardiac Rehabilitation.* University Park Press, Baltimore.

EXERCISE 6

Name _____

Laboratory Section _____ Date _____

RESULTS AND CONCLUSIONS

Standard EKG

Do the amplitudes of the P waves from the three standard leads follow Einthoven's law? Do the amplitudes of the QRS complexes from the three standard leads follow Einthoven's law?

Evaluate your EKG on lead II by determining the durations and amplitudes of the cycle components listed below and comparing them with the standard figures cited in the text of this exercise. Interpret any significant deviations that you find.

	Duration	Amplitude	Comments
P wave			
QRS complex			
PR interval			
T wave			
PR segment			
ST segment			
QT interval			

Exercise 6] Results and Conclusions

What is the number of heartbeats per minute? Is the heart rate regular or is arrhythmia present? Does the heart rate change during inspiration and/or expiration?

What is the duration of electrical systole? What fraction of the heart cycle is spent resting between beats?

EKG Under Various Experimental Conditions

Record your observations concerning the effects of moderate exercise, cigarette smoking, and various other experimental procedures on the EKG recording. How can you interpret those findings? Be certain to discuss any unusual or unexpected features of your EKG.

Materials

bicycle ergometer (see Note)
stopwatch

Note: A suitable instrument may be available from a physical education department. The essential point is to be able to adjust the ergometer to various known work loads. The Quinton-Monark model 850, an extremely useful and moderately priced ergometer, is available from Quinton Instruments, 3051 44th Avenue W., Seattle, WA 98199.

Procedures

1. Familiarize yourself thoroughly with the construction and operation of the ergometer. Detailed instructions are provided with the instrument.
2. The selection of proper work loads is an important part of the exercise test. Active, well-trained individuals have little if any chance of overstraining themselves. Subjects whom one would expect to have a lower working capacity (completely untrained, older, or more delicate individuals) should exercise at moderate levels; an initial work load of 300 kg-m/minute is perfectly adequate for these individuals. Adjust the work load on the ergometer to 300–450 kg-m/minute and have the subject pedal for 6 minutes. Record the heart rate by palpation of the radial (or carotid) artery for the last 30 seconds or 1 minute of the ride. The pulse taker should practice his task before the actual test is conducted.
3. Increase the work load to 600–750 kg-m/minute and have the subject pedal for an additional 6 minutes. Record the heart rate for the last minute.
4. The test must be stopped immediately if the subject experiences shortness of breath, stitch in the side, or any pressure or pain in the chest region. Under the conditions of moderate exercise recommended here, there is no reason whatsoever to expect any difficulties.
5. The exercise test outlined here can be used for an objective measure of circulatory efficiency as part of a program of physical conditioning. According to Astrand [1965] effective training is conducted by performing relatively heavy work for a period of 4 minutes, followed by a 4 minute rest, 4 minutes of exercise, etc., for a total of 30 minutes, several times a week. Progress is assessed periodically with the graded exercise test.

STUDY QUESTIONS

1. According to data presented in Figure 7–2, the well-conditioned individual has a heart rate of 122 at a work load of 900 kg-m/minute, while the poorly conditioned subject shows a heart rate of 164 in performing exactly the same amount of work. Does this mean that the cardiac output (i.e., the total amount of blood pumped per unit time) is approximately one third higher in the unconditioned subject (assuming that the two men are the same height and weight)? How would one investigate this experimentally? What is meant by stroke volume and how is it studied? What are the effects of athletic conditioning on heart rate and stroke volume?

2. What is the PWC_{170} for each of the individuals in Figure 7-2? An individual underwent a program of rigorous training on the bicycle ergometer over a period of 3½ months. Data for his performance at selected intervals during the training program are listed below. What are the effects of training on the PWC_{170} of this individual?

Heart Rate at Steady State During Work Tests on the Bicycle Ergometer

Time	Heart Rate	
	At 900 kg-m/minute	At 600 kg-m/minute
0	168	140
20 days	148	119
1 month	137	106
2 months	135	108
3½ months	135	108

3. What is a treadmill and how does it operate? What are some of the parameters that can be varied in using the treadmill in graded exercise testing? How is the device used in cardiac rehabilitation?

REFERENCES

Adams, F. H., et al. 1961. The physical working capacity of normal school children. *Pediatrics, 28*:243.

Astrand, P.-O. 1965. *Work Tests with the Bicycle Ergometer.* Monark-Crescent AB, Varberg, Sweden. (Available from Quinton Instruments.)

Astrand, P.-O., and I. Rhyming. 1954. A nomogram for calculation of aerobic capacity (physical fitness) from pulse rate during submaximal work. *J. Appl. Physiol., 7*:218.

Astrand, P.-O., and K. Rodahl. 1970. *Textbook of Work Physiology.* McGraw-Hill, New York.

Bevegard, B. S., and J. T. Shepherd. 1967. Regulation of circulation during exercise in man. *Physiol. Rev., 47*:178.

de Vries, H. A. 1971. *Physiology of Exercise for Physical Education and Athletics.* Wm. C. Brown Co., Dubuque, IA.

de Vries, H. A. 1971. *Laboratory Experiments in Physiology of Exercise.* Wm. C. Brown Co., Dubuque, IA.

de Vries, H. A., and C. E. Klafs. 1965. Prediction of maximal O_2 intake from submaximal tests. *J. Sports Med. Phys. Fit., 5*:207.

Naughton, J., and H. K. Hellerstein (eds.). 1973. *Exercise Testing and Exercise Training in Coronary Heart Disease.* Academic Press, New York.

Schneider, E. C. 1920. A cardiovascular rating as a measure of physical fatigue and efficiency. *J. Am. Med. Assoc., 74*:1507.

Schneider, E. C., and D. Truesdell. 1922. A statistical study of the pulse rate, and the arterial blood pressures in recumbency, standing and after a standard exercise. *J. Physiol., 57*:429.

Wilson, P. K. (ed.). 1975. *Adult Fitness and Cardiac Rehabilitation.* University Park Press, Baltimore.

EXERCISE 7

Name _____

Laboratory Section _____ Date _____

RESULTS AND CONCLUSIONS

Schneider Test

    Reclining pulse rate_____

    Pulse rate increase on standing_____

    Standing pulse rate_____

    Pulse rate increase immediately after exercise_____

    Return of pulse rate to normal after exercise_____

    Systolic pressure, standing, compared with reclining_____

Using the Schneider table (Table 7-1), compute the index. 18 is a perfect score; 14-16 is considered good; 8-13 is fair; 7 or less is poor. Be certain to take note of the physical and emotional status of the subjects, including the following data: amount of sleep, amount of physical activity, does subject smoke, time since the last meal, presence of any pathological conditions, anxieties or emotional problems, etc.

    Schneider Index_____

Harvard Step Test

    Duration of effort_____

    Heart beats from 1 minute to 1½ minutes in recovery_____

Exercise 7] *Results and Conclusions*

Physical Fitness Index_____

Collect and analyze the class data for this test. How do the results compare with the Schneider test?

Bicycle Ergometry

Work load #1_____Heart rate_____

Work load #2_____Heart rate_____

Plot your two experimental points on the graph in Figure 7-2, and extrapolate the line to a heart rate of 170. What work load could you sustain at this heart rate? This is the PWC_{170}.

PWC_{170} _____

Use the Astrand nomogram to obtain the maximum oxygen consumption value (expressed as liters per minute) of which you are capable. How does this compare with your Basal Metabolic Rate? (See Exercise 11). The Astrand nomogram may be found in Astrand [1965] and in Astrand and Rodahl [1970].

SECTION III

Respiration

EXERCISE 8

Basic Aspects of Respiration

OBJECTIVES In this exercise we shall familiarize ourselves with basic anatomical and physiological features of the lungs and respiratory passageways. Elementary aspects of the respiratory process will be tested with a pneumograph and recorder, including (1) respiratory movements; (2) postural changes and respiratory rate; (3) the effect of reading aloud; (4) cold stimuli and respiration; (5) the effect of drinking; and (6) the effect of certain other experimental procedures. Interpretation of the results will assist in understanding various factors influencing the rate and depth of breathing.

The human respiratory system consists of the chief organs of breathing, the lungs; the pleural cavity within which expansion and contraction of the lungs take place; various air passageways through which air is conducted to and from the lungs; and the muscles of respiration, including the diaphragm, internal and external intercostals, and certain muscles of the abdomen. During ordinary quiet breathing 6-12 liters of air pass in and out of the lungs each minute. Oxygen is absorbed directly into the blood, carbon dioxide is eliminated. A wide variety of factors influences the rate and depth of breathing, and an understanding of these factors constitutes a fundamental topic in the study of human physiology.

Inspired air is warmed, filtered, and humidified in the upper respiratory passages, from which it passes through the pharyngeal region and larynx into the principal air passageway, the trachea. The trachea (or windpipe) is a short cylindrical tube, enclosed and protected by tough cartilaginous rings, that begins just under the larynx, extends downward 4 to 5 inches, and subdivides at its lower extremity into two main branches, the right and left bronchi. These bronchi, in turn, divide and subdivide to form the highly ramified "bronchial tree" (see Figure 8-1). Fine branches are called bronchioles; terminal branches are respiratory bronchioles. A cluster of tiny air sacs or alveoli adjoins each respiratory bronchiole and the absorption of oxygen into the blood takes place in the many tiny capillaries surrounding the alveoli. The alveolar sacs, of which there are thought to be 150 million in each lung, present a vast surface area (often estimated to be 30-40 square feet) to the inspired air, but only a small portion of this surface is actively engaged in respiratory gas exchange at any one time during normal, quiet breathing.

The lungs are paired, conically shaped organs lying within the pleural cavity. Membranous layers enclose and protect each lung. The outer layer lines the thoracic wall and is termed the parietal pleura; the inner layer, the visceral pleura, extends over the actual surface of the lungs. A small space between the two layers, the pleural cavity, contains a lubricating fluid secreted by the pleural membranes that prevents friction between the lungs and thoracic walls as the lungs expand and contract. Inflammation of the pleural membranes in the condition known as pleurisy causes considerable friction and may be extremely painful.

The right lung is divided into three lobes and is thicker and broader than the left lung. It is also shorter and somewhat elevated at the base because the liver is located just under the diaphragm on the right side. The left lung has two lobes and contains a concave region, the cardiac notch, within which the heart is located (see Figure 8-1).

Each lobe of the lung is subdivided into many smaller compartments called lobules. Each lobule contains an arteriole, a venule, a lymphatic vessel, and a respiratory bronchiole. Alveolar sacs connected to the respiratory bronchioles are each composed of an extremely thin layer of squamous epithelium and an elastic basement membrane. The surfaces of the tiny air sacs are covered with a fine meshwork of capillaries, and the diffusion of gases into and out of the blood occurs across the capillary and alveolar walls.

Because the intrathoracic pressure is ordinarily below that of the surrounding atmosphere, considerable reinforcement of the muscular walls of the chest, both anterior-posteriorly and laterally, must exist to keep the thorax from collapsing. This muscular support is provided by the bony elements of the chest, the sternum, the ribs, and the vertebral column itself. The sternum, or breastbone, is the large flat bone in the anterior wall of the chest. The lateral edges of the sternum form points of articulation with the upper seven pairs of ribs through their cartilaginous extensions. The lower five pairs of ribs (the so-called "floating ribs") articulate with costal cartilages of other ribs, or, in the case of the last two pairs, form no attachments at all. All twelve pairs of ribs attach posteriorly to the twelve thoracic vertebrae of the spinal column. Outward and upward movements of the rib cage during inspiration cause a substantial increase in the internal dimensions of the thoracic cavity, both in the anterior–posterior axis and laterally, as air is drawn into the lungs.

Several sets of muscles work in concert to bring about rhythmic expansion and contraction of the chest. The diaphragm, a flattened but distinctly dome-shaped sheet of muscle separating the thorax from the abdominal cavity, is considered to be the chief muscle of inspiration. Contraction of this muscle causes a downward and forward movement of the dome, a process that displaces the abdominal organs and causes the abdominal wall to protrude. The typical respiratory movement called abdominal breathing can best be observed in a reclining subject.

Contraction of the external intercostal muscles elevates the sternum and moves the anterior articulations of the ribs upward and outward (thoracic breathing). Coordinated interplay of the diaphragm and external intercostal muscles brings about the normal process of inspiration. Expiration is ordinarily passive in that relaxation of the diaphragm and external intercostals produces a decrease in thoracic volume. The active contraction of other muscles, however, particularly the internal intercostals and several muscles of the abdomen, often assists in lowering the sternum and rib cage and forcing the diaphragm upward into the thoracic cavity.

Many features of the normal respiratory process may be investigated by means of simple but meaningful experiments. The rhythmic expansion and contraction of the

Exercise 8]   Basic Aspects of Respiration

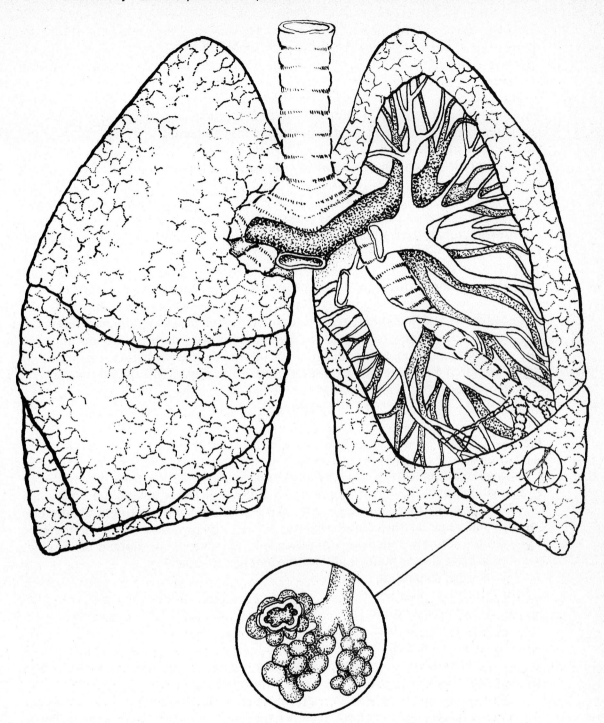

FIGURE 8-1   The human lungs. Identify and label the major anatomical features.

chest affords a convenient measure of basic respiratory processes and the various conditions that influence the rate and depth of breathing. In this exercise we shall study several elementary aspects of respiration using a simple device called a pneumograph.

MEASUREMENT OF RESPIRATORY MOVEMENTS

Materials

    pneumograph and tambour (1 setup per two to four students) (see Note)
    chart recorder or kymograph (1 setup per two to four students)
    8 ounce drinking glasses
    reading materials, including an anthology of poetry
    needles and thread
    ice cubes
    flexible tape measures (for measuring chest dimensions)
    disposable cardboard mouthpieces
    tape

Note: Many experimental setups are quite suitable. The equipment illustrated in Figure 8-2 makes use of the Harvard Apparatus Company's (Millis, MA) 8 inch chart mover equipment with pneumograph and tambour. Alternatively one can use the well-known Marey tambour and pneumograph recorded with a smoked drum, or ink-writing kymograph. The Narco Bio-Systems Physiograph can be equipped with a bellows- or impedance-type pneumograph. Other setups are possible.

Procedures

1. Observe movements of the chest during ordinary quiet breathing. Observe deep breathing as well. What is thoracic breathing? What is abdominal breathing? Have your partner make careful measurements of your chest (and abdominal) circumference at three positions—(a) second rib position, i.e., with the tape measure securely under the arm pits, (b) at the level of the xiphoid process (inferior tip of the sternum), and (c) at the umbilicus—after
    i. Normal inspiration.
    ii. Normal expiration.
    iii. Maximal inspiration.
    iv. Maximal expiration.
   Record the results on the data sheet.
2. Attach the pneumograph snugly at the chest position you have selected. Connect the pneumograph to the tambour at mid-inspiration and adjust the recording apparatus. If the pneumograph chain (line) slips, it may be secured with a cord around the subject's neck. Does an upward deflection of the pen correspond to inhalation or exhalation?
3. Record the rate of respiration for 2 or 3 minutes with the subject seated comfortably. It is essential that the subject not view the recording apparatus and that his attention not be focused on his own breathing processes. What is the respiratory rate? Repeat the determination with the subject in a reclining position. Finally, in a standing position. Are there differences?
4. Record quiet respiration for 1 minute with the subject seated. Have the subject read aloud for 30–45 seconds. Measure a 1 minute "recovery period." It is

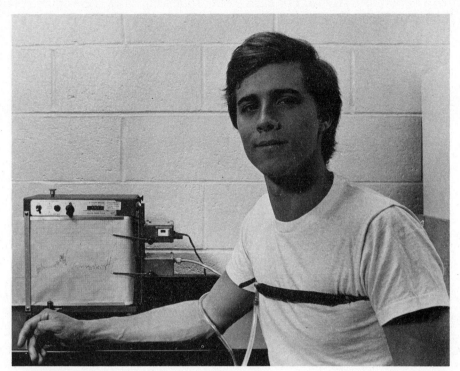

FIGURE 8-2 Pneumograph, tambour, and recorder for the measurement of respiratory movements.

especially interesting to have the subject read from an anthology of poetry, particularly rhythmic poetry.

5. Measure a control respiration rate for 1 minute and then stimulate the subject briefly by holding an ice cube for 2–3 seconds against each of the following:
   a. Palm of the hand.
   b. Earlobe (from the rear).
   c. Back of the neck.
   d. Center of the back.
   Record appropriate controls after each of the trials. The subject should not know where or when he is to be touched.
6. Record the chest movements while your partner drinks without stopping an 8 ounce glass of water. Include a 1 minute control before drinking and a 1 minute recovery period afterward.
7. Record the effects of the following on respiration:
   a. Voluntary coughing.
   b. Partial occlusion of the nasal passages by incomplete pinching of the nostrils.
   c. Concentrated mental effort such as performing a complicated arithmetic problem quickly or threading a needle.
8. *Hyperventilation and Involuntary Apnea.* Have the subject hyperventilate for 15–30 seconds in a sitting position. He should breathe as deeply and as rapidly as possible. Continue to record for at least 30 seconds after the test period. (Bear in mind that the subject may become dizzy from hyperventilation. Discontinue rapid breathing if this occurs, but keep recording).

9. *Voluntary Apnea.* For each of the following cases, determine how long the subject can hold his breath.
   a. After ten normal breaths following a *normal inspiration*.
   b. After a single *maximal inspiration*.
   c. After deep breathing for two minutes at twelve breaths per minute following *the last deep* inspiration.

STUDY QUESTIONS

1. What do the data summarized in Figure 8-3 indicate about the role of CO_2 in the control of breathing in man?

2. The respiratory minute volume of experimental subjects can be greatly increased by having them breathe gas mixtures high in CO_2 and low in O_2. A typical ventilatory volume for a subject in these experiments might average 75 liters/minute. The same subject, however, might show a respiratory volume of 150 liters/minute during extremely vigorous exercise. What does this finding indicate about the regulation of breathing by the pCO_2 and pO_2 of the blood?

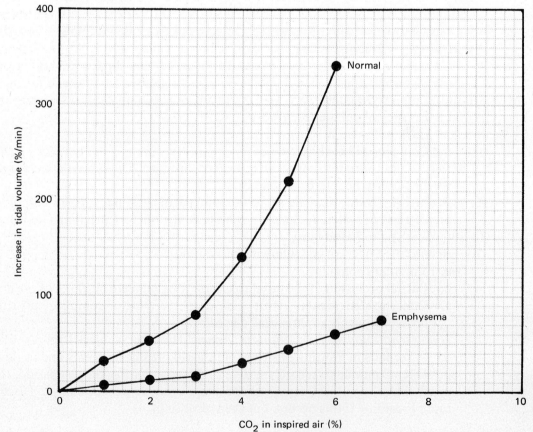

FIGURE 8-3 Increase in respiratory volume as the percentage of CO_2 in inspired air is increased. Note the relative tolerance to high proportions of CO_2 shown by the emphysematous patient. [*Adapted from* Human Respiration, A Programmed Course, *by Olof Lippold. W. H. Freeman and Company. Copyright © 1968.*]

3. Where in the nervous system is the control center for respiration located? Where is the inspiratory center and how does it function? Where is the expiratory center and how does it function? What is the pneumotaxic center? What is the actual experimental evidence for the statements you have made about each of these controls?

4. What is the Hering-Breuer reflex and what role does it play in the control of breathing? What is the carotid and aortic reflex and what part does it play? Again, cite specific experiments that underlie the statements you have made.

5. Describe the following processes in detail, each of which is closely related to and dependent upon respiration: coughing, sneezing, yawning, laughing, weeping, hiccoughing.

REFERENCES

Bouhuys, A. 1974. *Breathing: Physiology, Environment, and Lung Disease.* Grune and Stratton, New York.

Campbell, E. J. M, E. Agostoni, and J. Newton Davis (eds.). 1970. *The Respiratory Muscles; Mechanics and Neural Control.* 2nd ed. W. B. Saunders Co., Philadelphia.

Clements, J. A. 1962. Surface tension in the lungs. *Sci. Amer., 207*(6):120.

Comroe, J. H., Jr. 1966. The lung. *Sci. Amer., 214* (2):56.

Fenn, W. O. 1960. The mechanism of breathing. *Sci. Amer., 202* (1):138.

Grollman, S. 1974. *The Human Body: Its Structure and Function,* 3rd ed. Macmillan, New York.

Lippold, O. 1968. *Human Respiration: A Programmed Course.* W. H. Freeman and Co., San Francisco. (An excellent programmed text dealing with all aspects of respiration.)

Mead, J., et al. 1963. Factors limiting depth of breath of maximal inspiration in human subjects. *J. Appl. Physiol., 18*:295.

Morgan, T. E. 1971. Pulmonary surfactant. *N. Eng. J. Med., 284*:1185.

Nahas, G. G. (ed.). 1963. Regulation of respiration. *Ann. N. Y. Acad. Sci., 109*:411.

Patterson, J. L., Jr. 1974. Carotid bodies, breath holding and dyspnea. *N. Eng. J. Med., 290*:853.

Schaefer, K. E. 1958. Respiratory pattern and respiratory response to CO_2. *J. Appl. Physiol., 13*:1.

Smith, C. A. 1963. The first breath. *Sci. Amer., 209* (4):27.

Weibel, E. R. 1973. Morphological basis of alveolar capillary gas exchange. *Physiol. Rev., 53*:419.

EXERCISE 8

Name _____

Laboratory Section _____ Date _____

RESULTS AND CONCLUSIONS

Record the measurements of chest and abdominal circumferences below.

	Circumference at Level of		
	Second Rib	Xiphoid Process	Umbilicus
Normal inspiration			
Normal expiration			
Maximal inspiration			
Maximal expiration			

What is the best location for attachment of the pneumograph? Why?

What is the respiratory rate and relative tidal volume with the subject seated, standing, reclining? If any differences exist, how can you account for these?

Exercise 8] Results and Conclusions

How does reading aloud, especially from a poetry anthology, affect one's breathing? Conversely, how does one's breathing affect the reading?

Summarize the results you obtained after stimulating the subject with an ice cube. Can you account for the various observations?

What happens to respiration during drinking? What is the anatomical basis of this phenomenon? What is meant by "something going down the wrong pipe"? At what point(s) in the cycle of breathing does one drink?

Exercise 8] Results and Conclusions

Account for your observations concerning the effects of coughing, partial occlusion of the nasal passages, and concentrated mental effort on the rate and depth of breathing. What connection do your observations on partial occlusion have with pulmonary and bronchial disease?

What happens at the completion of the 30 second period of hyperventilation? Does involuntary apnea occur? If so, how long does it last?

Record the breath-holding times for the three conditions specified. What explanations can you advance for your observations? How could you test your suggestions?

Summarize your observations and draw pertinent conclusions concerning the effects of the various experiments on respiration. What specifically can you conclude about the overall regulation of breathing?

EXERCISE 9

Pressure and Volume Relationships in Human Respiration

OBJECTIVES In this exercise we shall acquaint ourselves with pressure and volume relationships as a basis for understanding respiratory processes in man. Three separate but related studies are to be carried out involving (1) Boyle's law and Charles' law; (2) a simple mechanical model of the human thorax; and (3) the measurement of various pulmonary volumes in the human respiratory system. Vital capacity measurements obtained in this exercise will be compared with those of a group of nonsmoking residents of Manitoba, Canada (a low-pollution area), using prediction equations derived from the literature.

BOYLE'S LAW: PRESSURE–VOLUME RELATIONSHIPS

The quantitative relationship between pressure and the volume of a gas was first described in the seventeenth century by Robert Boyle. Boyle's law states that, as a gas is compressed, the volume decreases in the same proportion as the pressure increases, provided the temperature is held constant. This law is usually written in the familiar form

$$P_1 \times V_1 = P_2 \times V_2$$

Consider the following example: A skin diver at sea level inhales deeply to give a total volume of 5 liters in his lungs, holds his breath, and dives to a depth of 33 feet. What is the volume of air in his lungs at this point? (Recall that 33 feet is the equivalent of 1 atm pressure). What is the gas volume at a depth of 66 feet? In making these calculations you may ignore temperature differences that would affect the final result.

A short film loop* provides the basis for a simple study of Boyle's law. In the experiment a sample of air is drawn into a 5 ml syringe, after which the outlet of the syringe is sealed. Standard weights are added to the barrel of the syringe to increase the pressure on the gas inside the chamber. As the pressure increases, the gas volume decreases and a series of measurements is made. The weight of the barrel (9.7 grams) and atmospheric pressure (14.7 lb/in.^{2} = 1030 grams/cm^2) exert pressure on the cross-

---

*Ealing film loop, *Boyle's Law* (see page 338).

sectional area of the syringe (0.74 cm²); this pressure is systematically increased as weights are added. View the film loop and tabulate the measurements on the data sheet at the end of this exercise.

CHARLES' LAW: TEMPERATURE–VOLUME RELATIONSHIPS

Charles' law states that a gas expands upon heating and contracts when cooled. A mathematical formulation that includes both Charles' and Boyle's laws is the familiar ideal gas law,

$$PV = nRT$$

P = pressure in atmospheres
V = volume in liters
n = number of moles of gas
R = universal gas constant (0.082 liter-atm/mole-°K)
T = absolute (Kelvin) temperature

Can you show that one mole of a gas occupies 22.4 liters at STP (standard temperature, i.e., 0°C = 273°K, and pressure, 1 atm)?

A second film loop* provides the basis for an introductory study of Charles' law. Two syringes, one containing helium and the second containing ordinary air, are mounted in a bath of peanut oil that is then heated over a considerable range of temperatures (25–125°C). As the bath is warmed, the gases expand. The point of this excercise is to measure the gas volumes at four selected temperatures: 25, 75, 100, and 125°C. A tiny stirrer is used to equilibrate the temperature of the oil bath. Carefully measure the gas volumes at each of the specified temperatures and tabulate your results on the data sheet.

THE MODEL LUNG

A mechanical model of the human thorax such as the one depicted in Figure 9-1 can be used to study pressure and volume changes within the "lung" and "pleural cavity" during a typical respiratory cycle. Although the volumes, pressures, and rates of air movement differ considerably between the model and actual human respiration, several general principles may be investigated with this simple device.

When the rubber diaphragm of the model is contracted (moved outward), the intrapleural space is increased, the pressure within this cavity decreases, and the lung inflates. As the intrapulmonary pressure is decreased, air is drawn into the lung. When the diaphragm is released, the intrapleural and intrapulmonary pressures increase, and air is forced out of the lung. Insertion of the diaphragm partway into the thorax simulates full expiration.

---

*Ealing film loop, *Thermal Expansion of Gases* (see page 338).

Exercise 9] **Pressure and Volume Relationships in Human Respiration**

Materials

Materials listed below are for one setup.

bell jar with a hole in the top
two-holed rubber stopper to fit the bell jar
"diaphragm" of fairly sturdy rubber (a dimestore punch ball is quite satisfactory)
small balloons
two glass T-tubes
straight glass tubing
thick walled rubber tubing
two simple U shaped manometers
two rulers (in millimeters)
two screw clamps
two stands and clamps (not shown in Figure 9-1)

Procedures

Mechanics of the Lung Model
1. Set up the apparatus as shown in Figure 9-1. The manometers should be firmly clamped to stands and the rulers secured to facilitate the taking of readings.
2. With the "trachea" open to the outside atmosphere, unclamp the tube leading into the pleural cavity; suck out by mouth some of the air until the balloon is

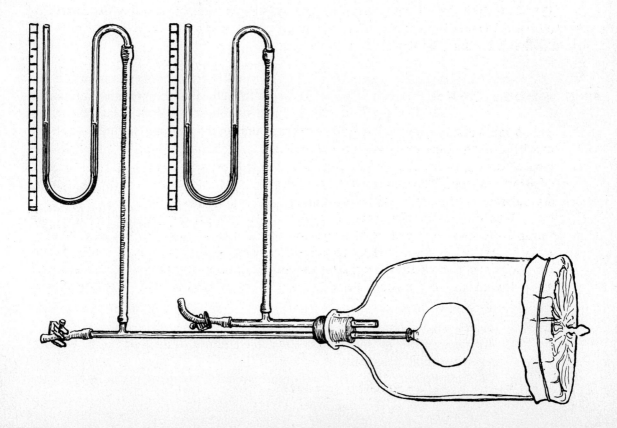

FIGURE 9-1 The lung-thorax model for assessment of pressure and volume changes.

partially inflated and fills about one quarter of the thoracic space. (This procedure simulates the human situation in which the intrapleural pressure is ordinarily below that of the surrounding atmosphere.)
3. Measure the pressures within the lung and pleural cavity. Pressure is recorded as the difference between mercury levels in the two arms of the manometer.
4. As you move the diaphragm in and out, observe and record the changes of pressure. Does the magnitude of the pressure changes vary if you increase the rate of respiration? If you increase the tidal volume?

Pressure Changes During the Respiratory Cycle
1. Clamp off the tube connecting the lung with the outside atmosphere and contract (pull out) the diaphragm as far as you reasonably can. Record the intrapulmonary and intrapleural pressures with the diaphragm in this position.
2. Admit small amounts of air into the balloon in a stepwise manner (five stages) until the intrapulmonary pressure equals atmospheric pressure. Record the two internal pressures at each step of this procedure.
3. After you have made these measurements, clamp off the trachea to the outside and release the diaphragm. Measure both pressures and then release the air in the lung to the outside in a series of five steps and measure the two pressures after each step. Record the readings in your data sheet.

Increased Resistance in the Air Passageway
1. As you slowly move the diaphragm back and forth, occlude the air passageway about half way by tightening the tracheal screw clamp. What effect does this have on the magnitude of pressure changes in the lung and pleural cavity? What do you have to do to inflate the lung to the same extent as with the air passageway open?

PULMONARY VOLUMES

During ordinary quiet breathing a relatively small volume of air passes in and out of the lungs with each breath. This air is a small fraction of the volume that can be inhaled and exhaled with deep breathing. In this exercise we shall measure various pulmonary volumes and relate one of these, the vital capacity, to individuals of matched age and height.

(1) *Vital capacity* (VC) = the amount of air that can be forcefully exhaled after maximal deep inhalation. Slow vital capacity (SVC) is the amount of air exhaled without regard to the rapidity of exhalation. Forced vital capacity (FVC) is the volume that can be exhaled as rapidly as possible. Differences between FVC and SVC are used to diagnose impairment of pulmonary function.

(2) *Tidal Volume* (TV) = the amount of air passing in and out of the lungs during respiration.

(3) *Inspiratory Reserve Volume* (IRV) = the total amount of air that can be inhaled beyond normal inspiration.

(4) *Expiratory Reserve Volume* (ERV) = the amount of air that can be exhaled beyond normal expiration.

(5) *Residual Volume* (RV) = the amount of air remaining in the lungs after maximal expiration.

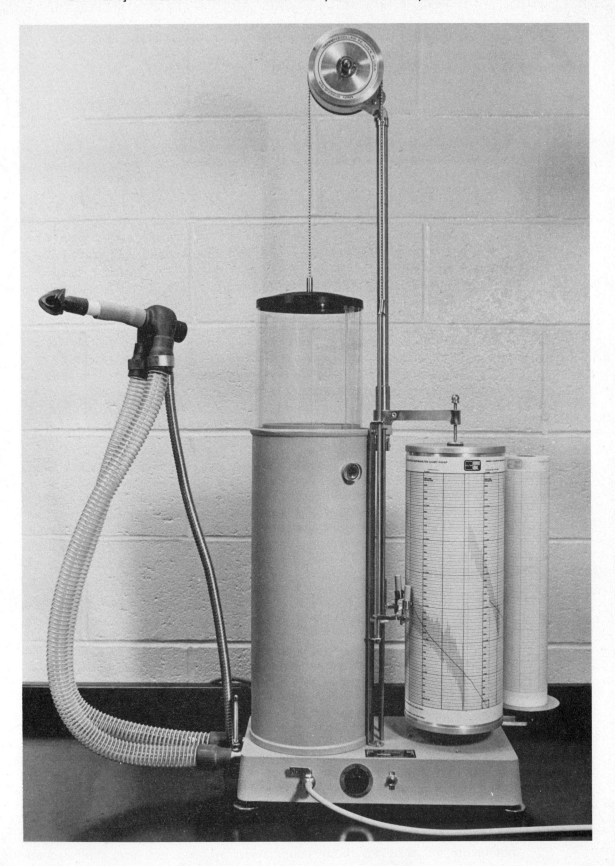

FIGURE 9-2 The Collins 9 liter respirometer.

Materials

Collins 9 liter respirometer or other suitable instrument (see Note)

Note: Available from Warren E. Collins, Inc., 220 Wood Road, Braintree, MA 02184.

Procedures

1. Familiarize yourself thoroughly with the basic design and the detailed operational mechanics of the respirometer. The device consists of a freely moving metal or plastic bell that floats in water inside a cylindrical container (see Figure 9-2). The breathing tubes connect directly to the inner chamber through one way rubber flutter valves. (One of the tubes leads into the bell; the other leads out. Can you determine which is which? Also, be certain to notice the construction of the plastic breathing tubes. Do they bear resemblance to any part of the human respiratory apparatus?).
Note the *respiration pen,* which gives a direct measure of the bell movement during inhalation and exhalation, and the *ventilograph pen,* which gives a summation of the inspired air (ventilograph reading × bell factor = the total volume of air inhaled per unit time).
2. Be certain that the rubber mouthpiece has been disinfected in 70% alcohol and thoroughly dried. Study the operation of the free-breathing valve to which the mouthpiece is attached. Be certain that the instrument contains active sodalime absorbant. Fill the respirometer with pure oxygen (through the stopcock at the base of the instrument).
3. The subject sits in a comfortable position and the rubber mouthpiece is inserted. Be certain that the breathing valve is open to the outside atmosphere. Place the noseclip firmly over the subject's nose and secure it in place. Adjust the pens to the surface of the recording paper and start the kymograph at slow speed (32 mm/minute). Close the breathing valve. Record the respiratory events for a short interval of quiet breathing. After 1 or 2 minutes, have the subject inhale as deeply as possible and exhale normally. This constitutes the IRV. After a short time have the subject exhale maximally (ERV). Finally, measure the VC by having the subject maximally inhale and exhale. (*Be absolutely certain that the respiration pen is more or less in the center of the kymograph, so that the maximal inhalation and exhalation can be properly recorded.*) It is advisable to make three separate measurements of the VC. Choose the highest value as your measurement. Figure 9-3 illustrates a typical spirometer tracing for these determinations.
4. Record the temperature of the gas in the respirometer by reading the thermometer at the base of the instrument. The temperature of the air within the lungs is 37°C (98.6°F), but the exhaled gas cools rapidly in the respirometer. (Is vapor condensation present in either of the spirometer breathing tubes?) Accordingly, corrections are necessary to change recorded volumes to actual volumes within the lungs (see Table 9-1). Multiply the measured VC times the correction factor for the particular spirometer temperature (e.g., 4.1 liters × 1.075 at 25°C) to obtain the VC at BTPS (body temperature, ambient pressure, saturated atmosphere).

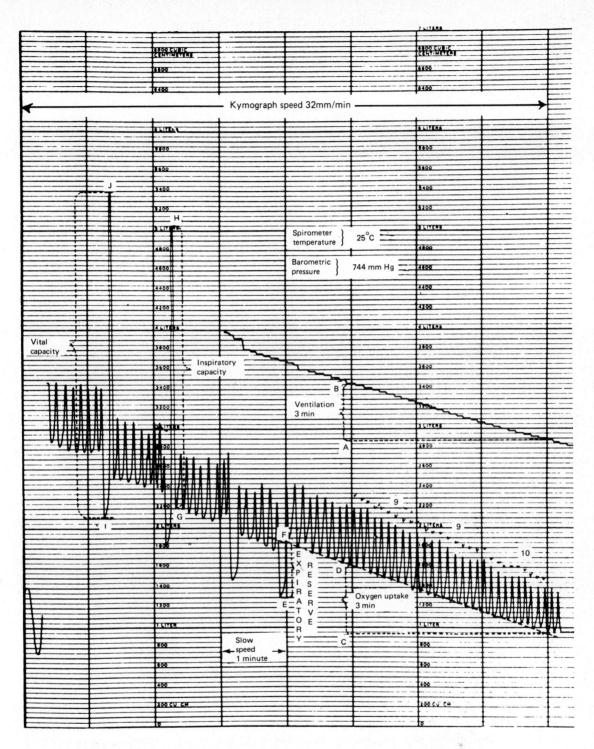

FIGURE 9-3 A typical spirometer tracing showing various respiratory parameters. The time scale reads from right to left; each vertical space represents 1 minute at a kymograph speed of 32 mm/minute. An *upward* deflection of the respiration pen (at the bottom of the illustration) records inspiration; *downward* deflection of the pen represents expiration. Expiratory reserve volume (ERV), inspiratory reserve volume (IRV), and vital capacity (VC) are shown. Lung ventilation (× 1/25) is represented in the upper tracing. [*Courtesy of Warren E. Collins, Inc.*]

TABLE 9-1 BTPS Correction Factors

Respirometer Temperature (°C)	BTPS Factor
20	1.102
21	1.096
22	1.091
23	1.085
24	1.080
25	1.075
26	1.068
27	1.063
28	1.057
29	1.045
30	1.039

STUDY QUESTIONS

1. Compare your vital capacity with a group of nonsmoking residents of Manitoba, Canada (a low-pollution area), as studied by Cherniack and Raber in 1972. Prediction equations based on statistical treatment of 859 men and 452 women in Manitoba are cited in the two formulas given below. Compute your predicted vital capacity and compare it with your measured value from this exercise.

   Men: $VC = 0.12102H - 0.01357A - 3.18373$ = predicted VC*
   Women: $VC = 0.07833H - 0.01539A - 1.04912$ = predicted VC

   H = height in inches
   A = age in years

2. Collect the class data and analyze them to see if a correlation exists between values for the students' percent of predicted vital capacity (BTPS) and cigarette smoking, athletic participation, experience with wind instruments and singing, and other variables. Summarize your results.

REFERENCES

Addington, W. W., et al. 1970. The association of cigarette smoking with respiratory symptoms and pulmonary function in a group of high school students. *J. Okla. State Med. Assoc.* (Nov. 1970), p. 225.

Beaver, W. L., and K. Wasserman. 1970. Tidal volume and respiratory rate changes at start and end of exercise. *J. Appl. Physiol.* 29:872.

Boren, A., et al. 1966. The veterans-administration-army cooperative study of pulmonary function. II: The lung volume and its sub-divisions in normal men. *Am. J. Med.,* 41:96.

Bouhuys, A. 1974. *Breathing: Physiology, Environment and Lung Disease.* Grune and Stratton, New York.

---

*One would obtain almost the same number by multiplying one's height times 0.121 rather than 1.12102, etc. To be consistent with the actual research papers from which these prediction equations have been drawn, however, the longer values are cited here, even though they obviously exceed levels of significance.

Cherniack, R. M., and M. B. Raber. 1972. Normal standards for ventilatory functions using an automated wedge spirometer. *Am. Rev. Resp. Disease, 106*:38. (Based on a study of nonsmoking residents of a low-pollution area: Manitoba, Canada.)

Collins, W. E., Inc. *Clinical Spirometry*. Braintree, MA. (Available from the company.)

Fry, D. L., and R. E. Hyatt. 1960. Pulmonary mechanics: A unified analysis of the relationship between pressure, volume and gas flow in the lungs of normal and diseased human subjects. *Am. J. Med., 29*:672.

Konno, K., and J. Mead. 1968. Static volume-pressure characteristics of the rib cage and abdomen. *J. Appl. Physiol., 24*:544.

Lippold, O. 1968. *Human Respiration: A Programmed Course*. W. H. Freeman and Co., San Francisco. (An excellent programmed text dealing with all aspects of respiration.)

Proctor, W. A., et al. 1968. The pressure surrounding the lungs. *Respir. Physiol., 5*: 130.

Weibel, E. R., and D. M. Gomez. 1962. Architecture of the human lung. *Science, 137*: 577.

EXERCISE 9

Name _____

Laboratory Section _____ Date _____

RESULTS AND CONCLUSIONS

Boyle's Law

Weight Added to Syringe (grams)	Volume of Air in Syringe (ml)	Barrel Weight (grams)	Pressure	
			Grams/0.74 cm²	Grams/cm²
0				
500				
1000				
2000				
3000				

Plot a graph of volume versus pressure. Is the relationship linear?

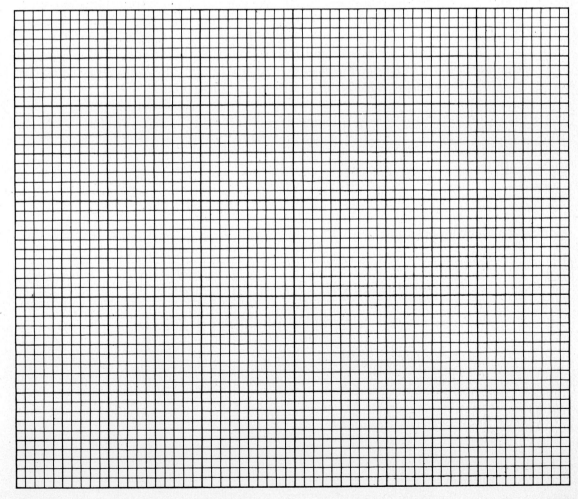

101

Exercise 9] Results and Conclusions

Is the product of $P \times V$ constant for each set of experimental conditions?

Charles' Law

Gas Volume		Temperature	
Helium	Air	°C	°K
		25	
		75	
		100	
		125	

Plot a graph of syringe volume versus temperature of the bath for helium and air. Two plots should be constructed; the first with temperature in degrees Celsius, the second in degrees Kelvin.

How do the results for helium and for air compare? What does this indicate about Charles' law and the behavior of gases? At what points do your extrapolated curves intersect the abscissa? What is the significance of these points?

Mechanics of the Model Lung

Experimental Condition	Intrapulmonary Pressure	Intrapleural Pressure
Resting position (lung partially inflated)		
Inspiration (slow breathing)		
Expiration (slow breathing)		
Inspiration (fast breathing)		
Expiration (fast breathing)		

Exercise 9] Results and Conclusions

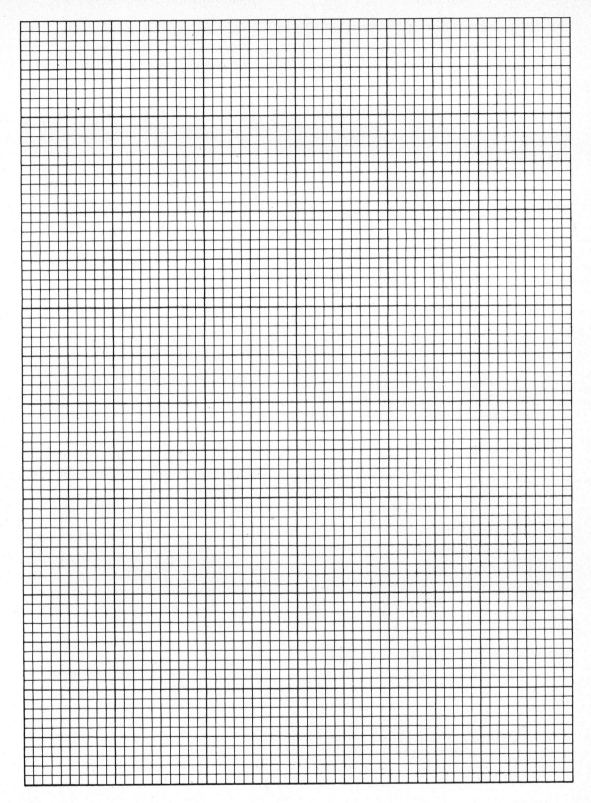

Exercise 9] Results and Conclusions

Pressure Changes During the Respiratory Cycle

	Intrapulmonary Pressure	Intrapleural Pressure
Full inspiration (trachea closed)		
Stage 1		
Stage 2		
Stage 3		
Stage 4		
Stage 5 (trachea open)		
Full expiration (trachea closed)		
Stage 1		
Stage 2		
Stage 3		
Stage 4		
Stage 5 (treachea open)		

Plot these data, using a solid line for intrapulmonary pressure and a dashed line for intrapleural pressure.

Increased Resistance in the Air Passageway

	Intrapulmonary Pressure	Intrapleural Pressure
Resting (trachea occluded $\times \frac{1}{2}$)		
Inspiration		
Expiration		

Exercise 9] Results and Conclusions

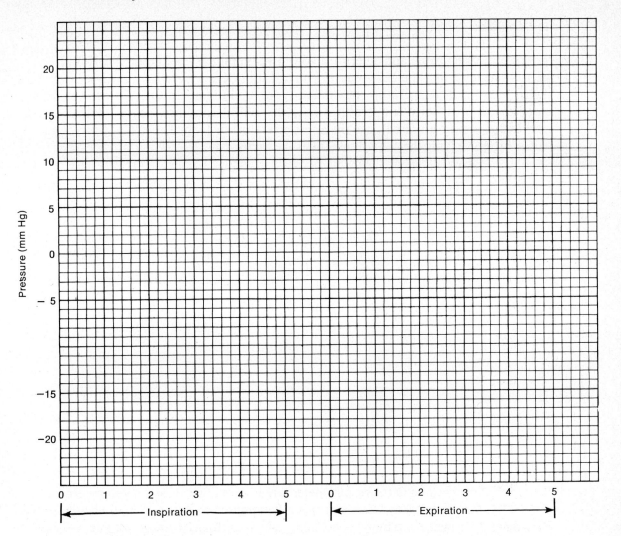

Exercise 9] Results and Conclusions

Pulmonary Volumes

Remove the chart paper and determine your own pulmonary volumes (TV, IRV, ERV, VC). Remember that the gas volumes you are considering have been measured in the respirometer at a temperature other than your body temperature. To correct these measurements to body temperature (i.e., the BTPS condition, body temperature, ambient pressure, saturated), use the appropriate factor from Table 9-1. Correct your values by multiplying the gas volume times the factor.

Is the volume measured in the respirometer larger or smaller than the volume that was present in your lungs? Why? Calculate also the minute volume, that is, the amount of air breathed during each minute of normal, quiet breathing. (Minute volume = TV × number of breaths per minute.)

EXERCISE 10

Pulmonary Function Tests

OBJECTIVES In this exercise we shall perform a series of pulmonary function tests, including (1) forced vital capacity (FVC) and forced expiratory volume (FEV); (2) maximal mid-expiratory flow rate (MMFR); (3) maximum breathing capacity (MBC); and (4) the match extinguishing test (MET). Class data will be used to study possible relationships between cigarette smoking, athletic participation, experience with singing and wind instruments, etc., and the functional status of the lungs. The data will also be analyzed using prediction equations derived from the literature.

Spirometric determination of the various lung volumes gives basic information about pulmonary compartments, but affords very little understanding of the actual functioning of the lungs. A more significant test is to measure the rate of air movement in and out of the lungs. In recent years various pulmonary function tests, making use of the Collins respirometer or other instruments of a similar kind, have been devised to analyze ventilatory function. Interpretation of the spirogram provides information about lung elasticity, respiratory musculature, and the status of the air passageways. Several of the simpler pulmonary function tests are presented in this exercise.

Although the measurement of static lung volumes is relatively unimportant in clinical diagnosis, it is nevertheless true that vital capacity measurements may sometimes be helpful in assessing pathological conditions or recovery from disease. For example, certain forms of heart disease produce blood congestion in the pulmonary capillaries, edema, and a consequent reduction in the vital capacity. As recovery proceeds, a corresponding increase in vital capacity is seen. Vital capacity may decrease substantially in polio because of paralysis of respiratory muscles.

FORCED VITAL CAPACITY AND FORCED EXPIRATORY VOLUME

Forced vital capacity, FVC, measures vital capacity under conditions where the subject exhales as vigorously as possible. Comparison of this value with ordinary vital capacity (measured without regard to the rate of expiration) is an important diagnostic test for several forms of obstructive pulmonary disease. In emphysema, for example, the alveolar walls become rigid and lose much of their elastic recoil. Upon inspiration the alveoli expand more or less normally, but upon expiration alveolar contraction (and lung deflation) are delayed. This leads to increased pressure within the thoracic cavity, collapse of many respiratory bronchioles, and trapping of air within the lungs. This is seen clinically as a significant difference between VC and FVC. A forced vital capacity less than 80% of slow vital capacity is usually taken as an indication of ventilatory impairment. The numerical difference between the two values is a measure of trapped air.

Forced expiratory volume (FEV) tests the kinetics of a single breath by measuring the velocity with which air can be expelled from the lungs at specifically timed intervals. $FEV_{1.0}$ is that volume of air forcefully exhaled in 1 second, $FEV_{2.0}$ is the volume expired in 2 seconds, and so on. A healthy person can exhale approximately 83% of his vital capacity in 1 second, 94% in 2 seconds, 97% in 3 seconds. More exact prediction equations are provided at the end of this exercise.

Materials

Collins 9 liter respirometer (or a suitable alternative)
70% alcohol for disinfection of the rubber mouthpiece
disposable mouthpiece connectors
noseclip
Collins VC timed interval ruler

Procedures

1. Remove the two flutter valves and the CO_2 absorbent container from the spirometer. Reposition the bell. Fill the bell with air and close the instrument to the outside atmosphere at the mouth valve. Have the subject stand in a comfortable position, insert the rubber mouthpiece so that the subject breathes room air, and attach the noseclip.
2. Vital capacity may be measured with the kymograph off. FVC is measured at 32 mm/minute. Adjust the breathing valve so that the subject is breathing from the spirometer, and measure slow vital capacity and forced vital capacity. Recall that 4 or 5 liters of air will be expired, so that the pen must be positioned properly on the chart to record this volume. You will probably want to make several recordings of vital capacity under the two conditions (slow and rapid expiration).
3. Forced expiratory volumes are best measured with the kymograph at top speed (1920 mm/minute). Correct positioning of the pen is essential for this test. Instruct the subject to inhale deeply and to exhale as vigorously as possible. Make three spearate recordings of timed vital capacity. Use the Collins VC timed interval ruler to measure $FEV_{1.0}$, $FEV_{2.0}$, and $FEV_{3.0}$. Record the results on the data sheet.

MAXIMAL MID-EXPIRATORY FLOW RATE (MMFR)

Mid-expiratory flow rate is measured directly from the graph of forced vital capacity. MMFR is defined as the rate of expiration during the middle half of the forced vital capacity measurement. The first quarter of the expiration is not considered because of a short time lag that usually occurs in FVC readings. In addition, some kinds of lung impairment (e.g., emphysema) do not affect the initial kinetics of the breath. The trapping of air in the lungs, a chief characteristic of emphysema, is not seen until the middle portion of the forced exhalation, and this results in a lowered MMFR value.

MAXIMUM BREATHING CAPACITY

Maximum breathing capacity (MBC) represents the amount of air moving in and out of the lungs in 1 minute. For practical purposes, the measurement is made over a period of 12 or 15 seconds and the value for 1 minute is calculated. This test constitutes an important assessment of the individual's capacity to ventilate his lungs. Figure 10-1 illustrates the spirometric record of this test.

Procedures

1. This determination can be made directly after measurements of FEV. The flutter valves and CO_2 absorbent are left out of the apparatus just as above.
2. Set the kymograph at medium speed (160 mm/minute) and after a short period of normal breathing, have the subject breathe as rapidly as possible for 15 seconds with deep inspirations and expirations. (Remember that the subject may become slightly dizzy from hyperventilation.)
3. Determine the rate of gas uptake from the record of the *ventilograph pen*. This factor is multiplied by the bell factor ($\times 25$) to give the actual volume of gas. Calculate the MBC (usually expressed as liters per minute) and record the results on the data sheet.

MATCH-EXTINGUISHING TEST (MET)

A very simple test of pulmonary function that is sometimes used by practicing physicians is to assess an individual's ability to blow out a lighted match. Carilli and Henderson [1964] refined this test under very careful laboratory conditions and showed that the match test correlated extremely well with $FEV_{1.0}$ and MBC.

In this exercise we shall carefully evaluate the ability to extinguish paper matches and correlate these data with the results of the previous assessments.

Materials

meter sticks (1 per group of students)
paper matches
ringstands and clamps (1 per group of students)
disposable mouthpieces
noseclips (1 per group of students)

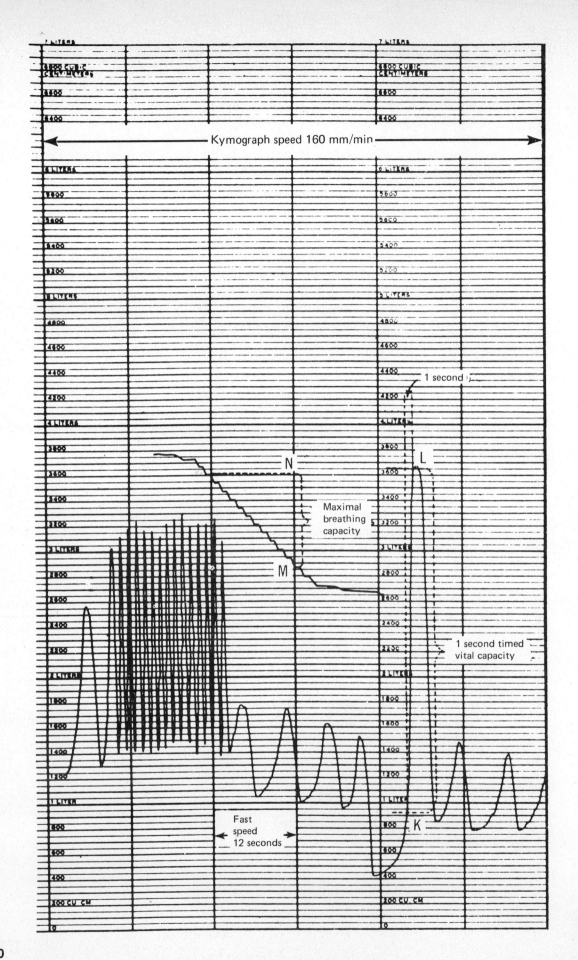

FIGURE 10-1 [opposite] A typical spirometric determination of timed vital capacity and maximum breathing capacity (MBC) at a kymograph speed of 160 mm/minute. Forced expiratory volume for 1 second ($FEV_{1.0}$) is depicted. Note that the *respiration pen* is used to determine timed vital capacity, and the *ventilograph pen* is used to measure MBC. The 15 rapid excursions of the respiration pen shown in the illustration correspond to the ventilographic recording of MBC. The two recordings are not found at corresponding positions because of the manner in which the pens are placed on the recording chart. [*Courtesy of Warren E. Collins, Inc.*]

Procedures

1. The details of the setup can best be worked out in one's own laboratory. Several suggestions will probably be helpful however. Orient a meterstick horizontally in a ringstand on the laboratory bench at head height (or a little lower). The disposable mouthpiece should be fixed in an immovable clamp and carefully aimed along the length of the stick. The match-blowing distance will probably vary in the class from about 60 to 130 cm; therefore, arrange the clamp holding the mouthpiece at a distance of 40 cm or so (measure it exactly) from the end of the meter stick.
2. Establish a set of criteria for the test and systematically evaluate each member of the class. Record the data.

STUDY QUESTIONS

1. Consider the following results from a series of pulmonary function tests administered to two adult men.

Case #1

	Measured	*Predicted*	*Percent of Predicted*
VC	4.5 liters	4.9 liters	_____ %
FVC	3.8 liters	4.9 liters	_____ %
$FEV_{1.0}$	2.7 liters	3.8 liters	_____ %
MMFR	160 liters/min	267 liters/min	_____ %

Case #2

	Measured	*Predicted*	*Percent of Predicted*
VC	3.3 liters	4.1 liters	_____ %
FVC	2.8 liters	4.1 liters	_____ %
$FEV_{1.0}$	1.8 liters	3.1 liters	_____ %
MMFR	72 liters/min	180 liters/min	_____ %

Calculate the "percent of predicted" for each of the tests. The men inhaled atomized spray containing a bronchodilatory substance and the pulmonary tests were repeated. In case 1, FVC, $FEV_{1.0}$ and MMFR each improved about 20% after administration of the bronchodilator. The bronchodilator produced no improvement in the tests in case 2. What can you conclude about the pulmonary status of each of these individuals?

2. Emphysema causes various types of cell and tissue damage in the lungs. These include a loss of elastic recoil of the alveolar sacs, partial destruction of alveolar tissue, thickening and distortion of capillary membranes, and partial collapse of terminal bronchioles with a consequent trapping of air in the lungs. How would these changes affect each of the following?
    (a) Ventilation in the lungs.
    (b) The diffusing capacity of gases in and out of the blood.
    (c) The oxygen and carbon dioxide content of the blood.
    (d) Acidity of the blood.
    (e) Blood pressure.
    (f) Intrathoracic pressure.

3. Many clinical studies have been carried out to evaluate the effects on the human respiratory system of acute and chronic exposure to various toxic agents, such as tobacco smoke, opiates, and marijuana products. Study one of the references listed at the end of this exercise dealing with these studies and summarize the findings. Be certain you understand the actual diagnostic procedures and pulmonary function tests that were used to assess respiratory impairment in the study you review.

REFERENCES

Auerbach, O., et al. 1972. Relation of smoking and age to emphysema. *N. Eng. J. Med., 286*:853.

Avery, M. E., N.-S. Wang, and H. W. Taeusch, Jr. 1973. The lung of the newborn infant. *Sci. Amer., 228*(4):74.

Bouhuys, A. 1974. *Breathing: Physiology, Environment and Lung Disease.* Grune and Stratton, New York.

Carilli, A. D., and J. R. Henderson. 1964. Estimation of ventilatory function by blowing out a match. *Amer. Rev. Resp. Dis., 89*:680.

Collins, W. E., Company. *Clinical Spirometry.* Braintree, MA.

Da Costa, J. L., et al. 1971. Lung disease with chronic obstruction in opium smokers in Singapore: Clinical, electrocardiological, radiological, functional and pathological features. *Thorax, 26*:555.

Dalhamn, T., and R. Rylander. 1963. Ciliostatic action of smoke from filter-tipped and non-tipped cigarettes. *Nature, 201*:401.

Gaensler, E. A., and G. W. Wright. 1966. Evaluation of respiratory impairment. *Arch. Environ. Health, 12*:146.

Leuallen, E. C., and W. S. Fowler. 1955. Maximum mid-expiratory flow. *Amer. Rev. Tuber. and Pulmon. Dis., 72*:783.

Lord, G. P., et al. 1969. The maximum expiratory flow-volume in the evaluation of patients with lung disease. *Amer. J. Med., 46*:72.

Motley, H. L. 1953. Use of pulmonary function tests for disability appraisal, including evaluation standards in chronic pulmonary disease. *Dis. of Chest, 24*:379.

Seely, J. E., et al. 1971. Cigarette smoking: Objective evidence for lung damage in teen-agers. *Science, 174*:741.

Tashkin, D. P., et al. 1973. Acute pulmonary physiologic effects of smoked marijuana and oral Δ-9-THC in young men. *N. Eng. J. Med., 289*:336.

Tennant, F. S., et al. 1971. Medical manifestations associated with hashish. *J. Amer. Med. Assoc., 216*:1965.

Tysinger, D. S., Jr. 1973. *The Clinical Physics and Physiology of Chronic Lung Disease, Inhalation Therapy and Pulmonary Function Testing.* Charles C Thomas, Springfield, IL.

EXERCISE 10

Name _____

Laboratory Section _____   Date _____

RESULTS AND CONCLUSIONS

Forced Vital Capacity

	Measured	Predicted	Percent of Predicted
Vital capacity (BTPS*)	_____ liters	_____ liters	_____ %
Forced vital capacity (BTPS)	_____ liters	_____ liters	_____ %

Prediction values for VC and FVC are identical and can be obtained from the following equations.

Men: $VC = 0.12102H - 0.01357A - 3.18373$
Women: $VC = 0.07833H - 0.01539A - 1.04912$

H = height in inches
A = age in years

Forced Expiratory Volume

	Measured	Predicted	Percent of VC
$FEV_{1.0}$ (BTPS*)	_____ liters	_____ liters	_____ %
$FEV_{2.0}$ (BTPS)	_____ liters	–	_____ %
$FEV_{3.0}$ (BTPS)	_____ liters	–	_____ %

Prediction values for $FEV_{1.0}$ can be obtained from the following regression equations.

Men: $FEV_{1.0} = 0.09107H - 0.02320A - 1.50723$
Women: $FEV_{1.0} = 0.06029H - 0.01936A - 0.18693$

H = height in inches
A = age in years

Prediction values for MMFR may be obtained by consulting the Collins nomogram (catalog number P-448).

*See Exercise 9 for an explanation of the BTPS correction.

Prediction values for MBC are calculated from the following regression equations.

Men: $\quad\quad\quad$ MBC = $(86.5 - 0.552A) \times m^2$
Women: $\quad\quad$ MBC = $(71.3 - 0.474A) \times m^2$

A = age in years
m^2 = body surface area (see Appendix A)

Match-Extinguishing Test

Greatest distance at which match is extinguished_____cm. Correlation diagrams can be drawn for the MET and any or all of the previous assessments (VC and FVC, $FEV_{1.0}$, MBC, MMFR). Students trained in statistical analysis will be able to apply regression analysis to the data. Collect the class data for the various tests and analyze them as suggested. (Carilli and Henderson [1964] report that MBC = 126 liters/minute corresponds to MET = 100 cm.)

Analysis of Class Data

Collect and analyze the class data to determine if there exists a correlation between the results of the various pulmonary function tests and cigarette smoking, athletic participation, experience with singing and wind instruments, and any other parameters you think are important.

EXERCISE 11

Indirect Measurement of Metabolic Rates

OBJECTIVES In this exercise we shall study the use of the respirometer in the indirect measurement of metabolic rates. Basal metabolism and metabolic rates following moderate exercise are to be determined. Sample calculations are provided to assist in quantitative evaluation.

The oxidation of organic materials by the human body liberates energy. Some of this is used to perform biological work, and much of it is released directly in the form of heat. Eventually all of the energy derived from biological oxidation is converted to heat. Accordingly, a direct measure of metabolic activity can be made from calorimetric determinations of the warmth given off by the human body. In the *direct method* the subject is placed in an insulated wholebody calorimeter and heat production is calculated from the temperature change of water flowing through the instrument. This procedure, although simple in conception, is ordinarily not performed in teaching laboratories because it requires elaborate and expensive equipment. Accordingly, an *indirect method* has been devised based on the rate of oxygen consumption. In this exercise we shall measure metabolic rates using the Collins 9 liter respirometer or a similar metabolator.

The complete oxidation of glucose according to the equation below yields 5.01 kcal of heat energy per liter of oxygen consumed.

$$C_6H_{12}O_6 + 6\,O_2 \longrightarrow 6\,CO_2 + 6\,H_2O + 673\text{ kcal}$$

Can you verify by calculation the figure of 5.01 kcal/liter of oxygen consumed? (*Hints:* How many liters of oxygen are needed to oxidize 1 mole of glucose? How much heat is produced in this reaction? Recall that 1 mole of a gas occupies 22.4 liters.)

The caloric equivalents for other foodstuffs are 5.06 kcal/liter of oxygen for carbohydrates (starches); 4.70 kcal/liter for fats; and 4.60 kcal/liter for proteins. For an average diet, the energy release is approximately 4.825 kcal/liter of oxygen consumed. Using this average *energy equivalent of oxygen,* you can make an indirect measure of energy expenditure in the human body from the rate of oxygen consumption. A simple determination of oxygen uptake in the respiration is related to heat production using this average energy equivalent. Table 11-1 summarizes the energy expenditure for a variety of human activities as measured by the indirect method.

TABLE 11-1 Energy Expenditure During Different Types of Activity for a 70 kg Man[a]

Form of Activity	kcal/hour
Awake (lying still)	77
Sitting at rest	100
Typewriting rapidly	140
Dressing	150
Walking (level grade at 2.6 miles/hour)	200
Bicycling (level grade at 5.5 miles/hour)	304
Walking (3% grade at 2.6 miles/hour)	357
Sawing wood or shoveling snow	480
Jogging (level grade at 5.3 miles/hour)	570
Rowing (20 strokes/minute)	828
Maximal activity (untrained individual)	1440

[a]From *Human Physiology–The Mechanisms of Body Function*, 2nd ed., by Sherman, Vander, and Luciano. Copyright © 1975 by McGraw-Hill, Inc. Used with permission of McGraw-Hill Book Company.

Since many factors influence the metabolic rate, it is essential to control as many of the variables as possible when comparing the metabolic rates of different individuals. The standard clinical test, called the *basal metabolic rate* (BMR), achieves this by rigorously standardizing the test conditions. The specific conditions prerequisite to the establishment of a basal metabolic state have been summarized by Guyton [1971]:

(1) The person must not have eaten any food for at least 12 hours because of the specific dynamic action of foods.
(2) The basal metabolic rate is determined after a night of restful sleep, for rest reduces the activity of the sympathetic nervous system and of other metabolic excitants to their minimal activity.
(3) No strenuous exercise is performed after the night of restful sleep, and the person must remain at complete rest in a reclining position for at least 30 minutes prior to actual determination of the metabolic rate. This is perhaps the most important of all the conditions for attaining the basal state because of the extreme effect of exercise on metabolism.
(4) All psychic and physical factors that cause excitement must be eliminated, and the subject must be made as comfortable as possible. These conditions, obviously, help to reduce the degree of sympathetic activity to as little as possible.
(5) The temperature of the air must be comfortable and be somewhere between the limits of 65 and 80°F. Below 65°F, the sympathetic nervous system becomes progressively more activated to help maintain body heat, and above 80°F, discomfort, sweating, and other factors increase the metabolic rate.

BMR is ordinarily expressed in kilocalories per hour per square meter of body surface (kcal/hr/m^2). Body surface area takes into account individual differences in height and weight and can be derived with a nomogram such as the one included in Appendix A. Average BMR values for men and women in various age groups are presented in Table 11-2.

TABLE 11-2 Average BMR Values[a] for Men and Women (kcal/hr/m^2 of body surface)[b]

Age	Males	Females	Age	Males	Females
15	45.3	39.6	35-39	39.2	35.8
16	44.7	38.5	40-44	38.3	35.3
17	43.7	37.4	45-49	37.8	35.0
18	42.9	37.3	50-54	37.2	34.5
19	42.1	37.2	55-59	36.6	34.1
20-24	41.0	36.9	60-64	36.0	33.8
25-29	40.3	36.6	65-69	35.3	33.4
30-34	39.8	36.2			

[a]Boothy and Sandiford Modification of DuBois Standards.
[b]From *Physiology and Biophysics*, 19th ed., edited by T. C. Ruch and H. D. Patton. Copyright © 1965 by W. B. Saunders Co. Used with permission of W. B. Saunders.

The percent deviation from normal is calculated as follows:

$$\text{BMR}(\pm\%) = \frac{\text{Predicted normal} - \text{Measured normal}}{\text{Predicted normal}} \times 100$$

If the measured value is less than normal, the BMR is designated by a minus sign; if the measured value is greater than normal, the BMR is reported as plus. Hypothyroid individuals may have a BMR of -15%; hyperthyroid individuals may have a BMR of $+40-50\%$.

Materials

Collins 9 liter respirometer (or a suitable metabolator)
70% alcohol for sterilization of mouthpieces
noseclip
laboratory barometer
tank of pure oxygen (designated for breathing purposes)
folding cot (or other accommodation for comfortable reclining)

Procedures

If possible, the subjects for this experiment should report to the laboratory in the morning and should not have eaten breakfast. If this schedule is not feasible, the students should report in the afternoon without having had lunch. The values obtained in this latter case will necessarily be somewhat higher and will not represent a true basal state.

The subjects must recline and rest quietly for 30 minutes prior to the determination. This step is essential and requires the cooperation of every member of the class.

1. Be certain that the one-way flutter valves and the CO_2 absorbent container are properly placed in the respirometer.

2. Fill the bell of the respirometer with pure oxygen and close the apparatus by opening the valve on the mouthpiece to the room air.
3. Insert the mouthpiece into the subject's mouth and apply the noseclip. This may be done during the last five minutes of the 30 minute rest period. The subject must remain in a reclining position both before and during the test.
4. Start the kymograph at slow speed (32 mm/minute) while the subject is still breathing room air. When the subject is breathing comfortably, turn the mouthpiece valve to connect the subject with the respirometer. Do this directly *after the subject has exhaled.*
5. Record the rate of oxygen consumption for 8 minutes. If time allows, repeat the procedure for a second 8 minute BMR determination. The subject should rest for several minutes between measurements.
6. Following completion of the BMR determination, the subject may perform moderate exercise (e.g., running in place for 5 minutes). After a 10 minute rest period, the post-exercise metabolic rate is measured. An alternative procedure is to measure the rate of respiration directly while the subject is exercising (e.g., marching in place at a cadence of 30 steps/minute).
7. Record the data in the laboratory report and calculate metabolic rates. The rate of oxygen consumption is measured from the slope of the respiration pen. Disregard the first 2 minutes or so of the reading during which time the subject is adjusting to breathing pure oxygen. If the rate of respiration is measured for less than 8 minutes, use a linear portion of the tracing after the initial 2 minute period.

STUDY QUESTIONS

1. The indirect method for the determination of metabolic rates probably contains several inaccuracies because of certain simplifying assumptions. Can you suggest what some of these errors might be?

2. What are the effects of adrenalin (epinephrine) and thyroxine on metabolism in the body? Where specifically do these hormones have their action and how are they thought to influence metabolic activity?

3. For what disease is the BMR measurement a helpful diagnostic tool? What additional tests can one perform to confirm preliminary diagnoses based on BMR?

4. Can you calculate the proportion of inhaled oxygen that is actually absorbed by the body during quiet breathing? (*Hint:* Inspect your spirometric chart for (1) the rate of pulmonary ventilation from the ventilograph pen and (2) the rate of oxygen uptake from the respiration pen.) Does this proportion change during moderate exercise?

5. A 70 kg man walks for 1 hour at a rate of 2.6 miles/hour. How much sugar would he have to consume to provide the energy for his walk? Consult Table 11-2. (The molecular weight of glucose is 180 daltons; 1 lb = 454 grams.) Suppose that the energy was derived from biological oxidation in the body and that body fat was the substrate for oxidation. How much weight would the man lose in 1 hour (aside from water loss)? What additional information do you need to answer the second part of this question?

6. The oxidation of one mole of glucose yields 673 kcal of energy. How much of this energy is actually captured in the form of ATP as glucose is degraded through the reactions of the glycolytic pathway and the Krebs cycle? Of the various ways in which the body "spends" ATP as "biological currency," which result in the more or less immediate release of heat energy? Can you cite at least one metabolic use of ATP in which a considerable portion of the energy is not released immediately as heat?

REFERENCES

Benzinger, T. H. 1961. The human thermostat. *Sci. Amer., 204*(1):134.

Boothby, W. M., J. Berkson, and H. L. Dunn. 1936. Studies of the energy of metabolism of normal individuals: A standard for basal metabolism, with a nomogram for clinical application. *Amer. J. Physiol., 116*(2):468.

Cabanac, M. 1975. Temperature regulation. *Ann. Rev. Physiol., 37*:415.

Jansky, L. 1973. Nonshivering thermogenesis and its thermoregulatory significance. *Biol. Rev., 48*:85.

Peters, J. P., and D. D. Van Slyke. 1946. *Quantitative Clinical Chemistry*. Williams and Wilkins, Baltimore. (An excellent discussion of indirect calorimetry is to be found here.)

Tepperman, J. 1968. *Metabolic and Endocrine Physiology,* 2nd ed. Year Book Medical Publishers, Chicago.

Wilkins, L. 1960. The thyroid gland. *Sci. Amer., 202*(3),119.

EXERCISE 11

Name _____

Laboratory Section _____ Date _____

RESULTS AND CONCLUSIONS

Subject

    Height _____ cm   _____ in.   Age _____ years

    Weight _____ kg   _____ lb   Sex _____

		Measured	
	Example	*Basal*	*Exercise*
Respirometer volume (liters) (a) Start of 6 minute test period (b) End of 6 minute test period	3.20 liters 1.90 liters		
(c) Oxygen consumed (liters)/6 minutes (c) = (a) − (b)	1.30 liters		
(d) Oxygen consumed/hour (d) = (c) × 10	13.0 liters		
(e) Barometric pressure (from laboratory barometer)	765 mm Hg		
(f) Temperature in respirometer	28°C		
(g) Water vapor pressure at temperature (f)	28.35 mm Hg		
(h) Volume of oxygen consumed at STPD (liters/hour) $(h) = \dfrac{(d) \times [(e) - (g)] \times 273}{[273 + (f)] \times 760}$	$\dfrac{13 \times (765 - 28.35) \times 273}{(273 + 28) \times 760}$		
(i) Kilocalories released/hour (i) = (h) × 4.825	57.2 kcal		
(j) Body surface area (meter^2) (see nomogram in Appendix A)	1.67 m^2 Height: 168 cm Weight: 60 kg		
(k) Metabolic rate (kcal/hr/m^2) (k) = (i)/(j)	34.2 kcal/hr/m^2		

(l) Predicted metabolic rate (see Table 11-2)	36.9 Age: 19 yrs Sex: female		
(m) Metabolic rate (percent deviation from predicted) $(m) = \dfrac{(l) - (k)}{(l)} \times 100$	BMR = 7½%		

SECTION IV

Nerve and Sensory Processes

EXERCISE 12

Reflexes in Man and the Nerve Impulse

OBJECTIVES In this exercise we shall study several types of reflexes in the human body affecting a variety of processes. We shall also view a film, *The Nerve Impulse,* that offers excellent reconstructions of classical experiments that have led to our present understanding of the nerve impulse.

Reflexive action constitutes a rapid and involuntary response to stimulation either from within or without the body, and involves the activity of the reflex arc, the most fundamental pathway in the nervous system. Tapping the tendon of the quadriceps muscle just below the knee causes extension of the lower leg; the presence of food within the small intestine induces secretion of digestive juices; a hot object that comes into contact with the skin causes rapid removal of the limb. Reflexes are unconscious activities in that we do not deliberately will them to occur; for the most part they involve the activity of the spinal cord independent of higher neural centers, although some do require the activity of the brain.

Reflexes are purposeful and may be categorized according to the functions they serve, such as postural reflexes, blood-pressure-regulating reflexes, flexor and extensor reflexes, etc. Reflexes may also be classified as visceral or somatic, depending upon the kind of organ that is stimulated to activity. Visceral reflexes affect heartbeat, breathing, and blood pressure; somatic reflexes involve the various skeletal muscles of the body. Finally, reflexes may be classified as to the location of the sensory receptors that initiate reflexive action. In an exteroceptive reflex, the receptor is at or near the surface of the body. In a visceroceptive reflex, the sensory element is located within a visceral organ or blood vessel. In a proprioceptive reflex, the receptor is within a muscle or tendon.

A typical reflex arc establishes connection between a receptor and an effector. In most cases the receptor is the distal end of a sensory neuron. The sensory element responds to stimulation by generating an action potential that is passed along to the spinal cord, where it may be transmitted farther or blocked. In the simplest case (axon reflex) the sensory impulse is passed directly to an effector, such as an arteriole. In other instances the sensory neuron synapses with a motor neuron and the inpulse passes along the motor pathway to an effector, such as a muscle or gland. This type of reflex arc, known as a two-neuron monosynaptic arc, is exemplified by the well-known knee jerk reflex. Most reflex arcs are rather more complex and involve at least three separate neurons. In this case, the sensory neuron synapses with associative neuron(s)

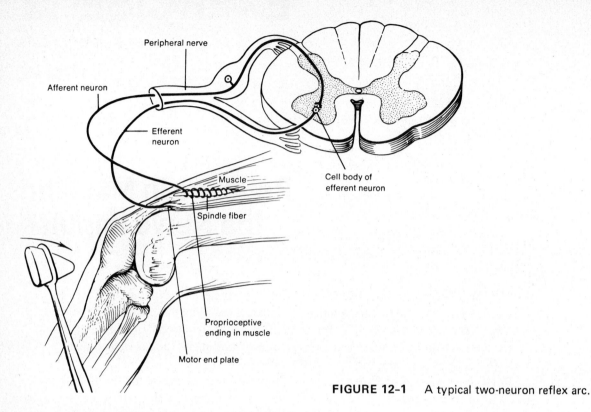

FIGURE 12-1 A typical two-neuron reflex arc.

that pass the impulse to motor elements. Figures 12-1 and 12-2 illustrate typical two- and three-neuron reflex arcs.

Examination of reflex status is an important diagnostic measure. In clinical terms reflexes are categorized as deep or superficial. Deep reflexes are all stretch (myotactic) reflexes such as those elicited by a sharp tap on the appropriate tendon or muscle to

FIGURE 12-2 A typical three-neuron reflex arc.

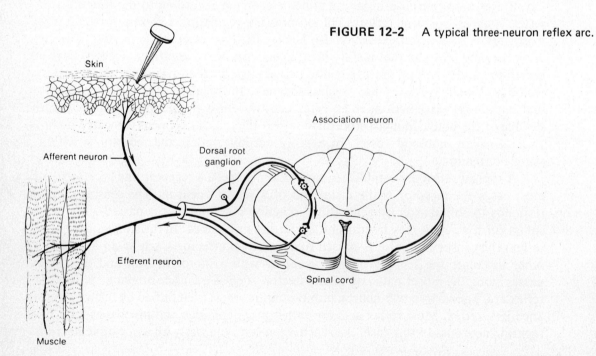

induce stretching of the muscle. Superficial or cutaneous reflexes are withdrawal reflexes induced by noxious or tactile stimuli. Neural disorders may affect reflex activity in any one of the following ways:

(1) The reflexes may be diminished or absent altogether (hyporeflexia).
(2) The reflexes may be hyperactive (hyperreflexia).
(3) The pattern of reflex activity may be altered in such a way that a new "pathological reflex" is present.

It must be emphasized that a wide variety of factors affects reflex activity so that proper interpretation of clinical findings requires a cautious approach. Preliminary diagnosis must be confirmed through repeated testing under a variety of circumstances. If reflex aberrations are found to be asymmetrical (if, for example, one limb behaves differently from the other), then the clinical interpretation is much firmer; in this case one part of the body, in effect, serves as a control for its opposite member.

REFLEX ACTIVITIES IN MAN

Materials

small flashlights
lens paper
rubber reflex hammers
drinking cups
swivel chair(s)

Procedures

Direct Light (Pupillary) Reflex
1. Observe the size of your partner's pupils in ordinary room light. Shine a fairly bright light onto his eye and note the size of the pupils. What happens? Remove the stimulus and again note the pupil size. What happens? Record your results.
2. What is the physiological significance of this reflex? Can you draw the exact neural pathway that is involved? What is the sensory element? What is the motor element? Which cranial nerve(s) is (are) involved? What is the effector?

Consensual Light Reflex
1. Ask your laboratory partner to hold his hand (or a book) vertically in front of his face to separate the right eye from the left. With a small flashlight illuminate one eye, but avoid shining light on the other. Note the size of the pupils in both eyes. Record your observations.
2. Repeat this procedure with the other eye. Have your partner cover one eye with his hand. What happens to the pupil of the uncovered eye? Describe the anatomical basis of this response.

Corneal Reflex
1. Carefully touch your partner's cornea with a piece of fine lens tissue. What happens? Is this reflex consensual? Record your observations.

2. Why is it called the corneal reflex? Explain how cranial nerves V and VII participate in this reflex.

Accommodation Reflex
1. Have your partner focus his sight on an object held by a third student at a distance of 15 or 20 feet. Carefully observe the size of the subject's pupils as the object is brought closer to his eyes. The student carrying the object should be careful not to block the source of illumination in the room, as the light intensity should remain constant. (Why?) Record your observations. What other changes are taking place in the eye as the focus is made shorter?

Knee Jerk (Patellar) Reflex
1. Have your laboratory partner sit comfortably in a chair with crossed legs. The top leg should hang freely. Tap the patellar tendon with the reflex hammer and note the response. Test both legs in succession.
2. Test the knee jerk reflex under the following conditions.
    a. While the subject is engaged in concentrated mental activity, such as threading a needle or mentally adding a long list of numbers.
    b. While the subject with interlocking fingers is pulling one hand against the other.
    c. After the subject has become fatigued by running around the building or up a flight of stairs.
    Record all your results.

Ankle Jerk (Achilles) Reflex
1. This reflex is tested with the subject in the same position as above. Tap the Achilles tendon with the reflex hammer and note the response. You may find it helpful to press gently but firmly against the subject's foot.
2. Repeat the test under conditions (a) and (b) above. How do the results compare?

Pharyngeal Reflex
1. Swallow as many times as you can in a 20-second interval.
2. Repeat the test, but take a small amount of water into your mouth each time to aid swallowing. How do the results compare? Is swallowing a reflex? Where are the sensory receptors located? What stimulates the receptors to initiate swallowing? What muscles are the effectors?

Red Skin Response
1. Rub the surface of your arm rather severely with a blunt object. What happens? Is this an example of a reflex? Explain.

Biceps Reflex
1. Flexion of the forearm may be achieved by tapping the tendon of the biceps brachii. The examiner places the thumb of his left hand firmly over the subject's tendon and strikes it (the thumb) with the reflex hammer. What happens? What type of reflex is this? What is the sensory receptor in this case? Which nerve(s) and spinal segment(s) are involved in this action?
2. A similar test can be conducted with the tendon of the triceps muscle, at a point slightly above the elbow.

Plantar Reflex

1. Have your partner remove a shoe and sock. Scratch the sole of his foot by moving a blunt object along the sole toward the toes. Flexion of the toes is a normal response. If the big toe extends upward and the smaller toes spread apart, a pathological condition is present. This abnormal reflex is known as the *Babinski sign* and is commonly held to indicate neural damage in the pyramidal tract of the spinal cord. Interestingly, children under the age of 18 months normally show the Babinski response, especially when they are asleep.

Vestibular Reflexes

1. Test static equilibrium in your partner in the following way. Have him stand quietly with the feet together and eyes open. Carefully note the amount of body movement that occurs. Now have the subject close his eyes and again note the body movements. Repeat these simple tests with the subject's feet aligned in a straight line facing forward. Record your observations. What conclusions can you draw?
2. Have your partner sit in a swivel chair with his feet off the floor, and rotate the chair rather rapidly. Note eye movements (nystagmus) as the chair is being rotated. Stop the chair abruptly and again note the eye movements. Record your results. Now have your partner close his eyes; repeat the experiment. He is to state when he perceives the motion to stop. Next rotate the chair and allow it gradually to slow down and ask the subject to state when he thinks the movement has ceased. Record the results.
3. With the subject seated and his eyes closed, rotate the chair for a period of 10 or 15 seconds. Stop the chair abruptly and ask the subject to stand. *Several students must be positioned around the chair to keep the subject from falling.* In which direction does the student start to fall? Repeat the test after rotating the swivel chair in the opposite direction. Record the results. How can you interpret these findings?

THE NERVE IMPULSE

The Nerve Impulse, * a highly useful instructional film, presents a detailed history of scientific investigation into the nature of nerve impulse conduction and transmission, starting with the classical experiments of Luigi Galvani in the 1780s on electricity and muscle contraction. The film illustrates, in chronological sequence, the work of Galvani; Emil duBois-Reymond's research into electrical potential and its relation to nerve impulse conduction; von Helmholtz's measurements of the rate of nerve impulse conduction; Julius Bernstein's studies on the role of ions; Ramon y Cajal's discovery of neural disconnections and the synapse; Otto Loewi's important discovery of the chemical nature of synaptic transmission; and the Hodgkin-Huxley experiments on the giant axon of the squid that led to the discovery of resting potential.

While viewing the film, try to make a mental summary of the main points of each experiment that is shown. Class discussion following the film should serve to elucidate points that are unclear and help to augment the historical perspective of the film.

---

*The Nerve Impulse, produced by the Encyclopedia Britannica Educational Corporation (see page 339).

STUDY QUESTIONS

1. Most nerves in the body are "mixed nerves," that is, they contain both sensory and motor elements. Are there any nerves that serve only a sensory function? Only a motor function? If so, what are they? How is it known that these nerves are not "mixed"?

2. Discuss in detail a visceroceptive reflex activity, that is, a reflex involving one of the abdominal organs. What is the sensory element and to what is it sensitive? What is the motor element and to what effector does it lead? What part of the nervous system is involved in this reflex?

3. In this exercise we studied two separate pupillary responses; one as a response to light, the second as a response to an object brought closer to the eyes from a distance. In certain pathological states the so-called Argyll-Robertson pupil is noted. What is this condition, and what is its underlying cause?

4. Reinforcement phenomena are often noted in the study of reflex activities (cf. your investigation of the patellar reflex). How is reinforcement to be explained?

5. Reflex activity on casual examination seems very simple. Closer study points to a more complex situation as shown by the following example. Limb extension as part of a reflex activity involves both the *contraction* of extensor muscles and the *relaxation* of flexors. The stimulation and inhibition are sometimes termed reciprocal innervation; the inhibitory activity is called Sherrington's inhibition. How is this phenomenon to be explained, especially since the central nervous system does not contain inhibitory fibers innervating the skeletal muscle fibers? Cite actual experimental evidence and try to diagram the processes you describe.

6. In tabes dorsalis (locomotor ataxia, tertiary syphilis), lesion formation begins in the dorsal root ganglia of the spinal cord. How does this affect reflex activity? Herniated vertebral discs and several forms of polyneuritis are known to compress both dorsal and ventral roots. How does this affect reflex activity? Cite one condition that produces hyperreflexia.

7. Outline the basic elements of the theory for nerve impulse propagation. What is the experimental evidence that underlies this theory?

REFERENCES

Adrian, R. H. 1974. *The Nerve Impulse.* Oxford University Press, London. (Available from Carolina Biological Supply Company.)

Beers, W. H., and F. Reich. 1970. Structure and activity of acetylcholine. *Nature, 228*:917.

Cohen, L. B. 1973. Changes in structure during action potential propagation and synaptic transmission. *Physiol. Rev., 53*:373.

Eccles, Sir J. 1965. The synapse. *Sci. Amer., 212*(1):56.

Fiorentino, M. R. (ed.), 1970. *Reflex Testing Method for Evaluating Central Nervous System Development.* Charles C Thomas, Springfield, IL.

Gage, P. W. 1976. Generation of end plate potentials. *Physiol. Rev., 56*:177.

Granit, R. 1972. *Mechanism Regulating the Discharge of Motorneurons.* Charles C Thomas, Springfield, IL.

Gray, E.G. 1973. *The Synapse.* Oxford University Press. (Available from Carolina Biological Supply Company.)

Hodgkin, A. L. 1964. *The Conduction of the Nervous Impulse.* Charles C Thomas, Springfield, IL.

Jacobson, H., and R. K. Hunt. 1973. The origins of nerve cell specificity. *Sci. Amer. 228*(2):26.

Katz, B. 1966. *Nerve, Muscle and Synapse.* McGraw-Hill, New York.

Katz, B. 1971. Quantal mechanism of neural transmitter release: Nobel Prize Lecture, 1970. *Science, 173*:123.

Krnjeric, K. 1974. Chemical nature of synaptic transmission in vertebrates. *Physiol. Rev., 54*:418.

Landowne, D., et al. 1975. Structure-function relationships in excitable membranes. *Ann. Rev. Physiol., 37*:485.

Lowenstein, W. R. 1960. Biological transducers. *Sci. Amer., 203*(2):98.

Thomas, R. C. 1972. Electrogenic sodium pump in nerve and muscle cells. *Physiol. Rev., 52*:563.

Truex, R. C., and M. B. Carpenter. 1969. *Human Neuroanatomy,* 6th ed. Williams and Wilkins, Baltimore.

EXERCISE 12

Name _____

Laboratory Section _____ Date _____

RESULTS AND CONCLUSIONS

Reflex Activities in Man

For each of the reflexes listed below indicate the type of reflex activity being studied and include a detailed description of the sensory element, the motor element, the effector(s) and the region(s) of the nervous system involved in the activity. In addition, provide answers to the several questions listed in the description of procedures.

(1) Direct light (pupillary) response

(2) Consensual light reflex

(3) Corneal reflex

(4) Accommodation reflex

(5) Knee jerk (patellar) reflex

(6) Ankle jerk (Achilles) reflex

(7) Pharyngeal reflex

(8) Red skin response

(9) Biceps reflex

(10) Plantar reflex

(11) Vestibular reflex

The Nerve Impulse

Summarize the material presented in the film dealing with each of the classical investigations on the nature of the nerve impulse. In reviewing the content of the film try to summarize the actual experimental procedures that were carried out, the observations that were made, and the conclusions that resulted from the experiments. Particular points to be considered in detail include

(1) The use of microelectrodes in measuring resting and action potentials in a nerve.
(2) Nerve fiber threshold and the all-or-none principle.
(3) The nature of the "refractory period."
(4) The response of a sensory nerve to increasing stimulus strength.
(5) The role of acetylcholine in synaptic transmission.

EXERCISE 13

Some Elements of Human Vision

OBJECTIVES In this exercise we shall acquaint ourselves with various anatomical features of the human eye and use this understanding as the basis for a systematic study of (1) binocular vision and convergence, (2) depth perception, (3) eye preference, (4) visual acuity, (5) properties of the lens as illustrated with the Ingersoll eye model, (6) accommodation, (7) pupillary responses, and (8) properties of the retina.

As has often been stated, the human eye as a kind of instrument bears a striking resemblance to the camera and similar optical devices. However, the analogy has been overextended, especially in many introductory texts, in that essential differences between eye and camera are often understated or disregarded. In this exercise, dealing with several basic properties of the human eye, the student should try to develop a feeling both for similarities and differences between human vision and various aspects of the photographic process. In addition, he should be keenly aware that, although there does exist a substantial body of knowledge concerning the neural pathways and attending processes involved in normal vision, there is no satisfactory theory of "how we see." Indeed, the process of vision, with its strong connections to consciousness itself, constitutes a formidable mystery that scientists of the future will have to ponder and fathom.

ANATOMY OF THE EYE

As a preliminary to the experimental work of this exercise, study the drawing of the eye (Figure 13-1) and identify the following structures: cornea, sclera, conjunctiva, choroid layer, ciliary muscle, suspensory ligaments, iris, pupil, aqueous humor, retina, fovea centralis, optic disc, optic nerve, and vitreous humor.

Carefully examine your laboratory partner's eyes and identify the pupil, iris, sclera, cornea, and eyelids. Can you see the lens? Can you discern the sclerocorneal junction? Where are the lacrimal (tear) glands? What observations could you make to verify the location of these glands? When a person's eyes redden, what is happening? Where? Make a careful drawing of your partner's eye.

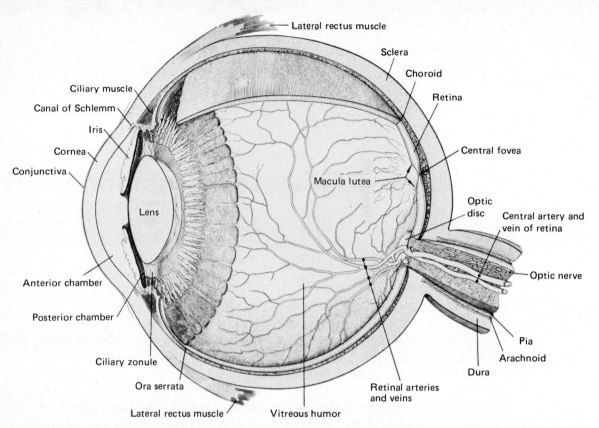

FIGURE 13-1 The human eye. Locate the various components as indicated in the text.

BINOCULAR VISION AND CONVERGENCE

When the two human eyes are focused on an object directly in front of an observer, each eye has its own particular field of vision that can be accurately charted with a perimeter. One of the signficant aspects of human vision is that the two separate visual fields overlap to a very great extent; this correspondence forms the area of binocular vision. Every single point in the field of binocular vision forms an image on the retinas of both eyes. A given object may be seen either as one object or as two, however, depending upon whether its image does or does not fall on *corresponding regions of the retina*. In most instances the two foveae serve as the corresponding points on the two retinas. The concept of corresponding points may be demonstrated in the following way. Look at an object in the room and then depress one eye very lightly by applying gentle pressure (over an eyelid) to the eye. What happens and how does this relate to the concept of corresponding points?

Quite obviously, when a person focuses on a very near object his two lines of sight converge on the object as his eyes are turned inward (i.e., are slightly crossed), so that the two images fall on the corresponding points. For viewing more distant objects, the two lines of sight also converge, but the angle between them is much more acute as the eyes are directed farther away. Can you diagram this?

In this part of the laboratory exercise we shall perform a simple but highly lucid demonstration of binocular vision and convergence in man.

Materials

10-foot lengths of string or cord
small one-hole wooden beads or rubber balls that can be threaded on the string

Procedures

1. Tie the string to the wall at eye level. Thread the bead on the string so that it moves freely along the length of the string.
2. Stand away from the wall with the string held taut from the wall to a point midway between your eyes. Focus your gaze directly on the bead. What do you see?
3. Your partner will move the bead away from you and toward you as you continue to focus on the bead. Describe what happens.

DEPTH PERCEPTION

With one eye closed it is possible, in viewing a particular scene, to obtain some impression of the spatial relationships involved, but one's perception of depth and the experience of three dimensions are substantially enhanced when both eyes are used. Indeed, binocular vision and convergence form the real basis of depth perception.

The Howard–Dolman apparatus commonly used in the assessment of this capacity contains a short vertical rod mounted in a rectangular box. A second movable rod parallel to the first is manipulated with a cord at a distance of 20 feet until the two rods appear to be parallel. The error in aligning the rods is measured for a series of trials. Figure 13–2 depicts the apparatus.

FIGURE 13-2 The Howard–Dolman apparatus for the measurement of depth perception.

Materials

Howard–Dolman apparatus (see Note)
measuring tape

Note: Available from Stoelting Co., 1350 South Kostner Avenue, Chicago, IL 60623.

Procedures

1. Set the depth perception apparatus on a laboratory bench and place a chair at a measured distance of 20 feet from the instrument.
2. Manipulate the cord until you think the vertical rods are parallel. Your laboratory partner will evaluate your attempt. Perform the procedure four more times and average the results. Repeat the test with one eye covered.

DETERMINATION OF EYE PREFERENCE

Although binocular vision is a cardinal feature of human sight, ordinarily one eye is the preferred or dominant eye. In general, a right-handed person is also right-eyed, and vice versa, but the correlation is not absolute.

Procedures

1. Cut a 1 cm hole in a 3 × 5 inch index card.
2. Place a penny on the floor. From a standing position with both eyes open locate the penny through the hole with the card held at arm's length.
3. Close each eye alternately and determine which eye you used in viewing the coin.

VISUAL ACUITY

The ability to resolve visual images (visual acuity) is tested with the well-known Snellen chart. A normally sighted person can see the "E" on the first line of the chart at a distance of 200 feet. Other lines correspond to shorter distances. Visual acuity (V) is defined as follows: $V = d/D$, where d is the distance at which particular letters can be read by the subject being tested, and D the distance at which these letters can be resolved by a normal eye. For example, a person who must be only 20 feet away to read a line on the Snellen chart that normally can be read at 40 feet has visual acuity in that eye of 20/40.

Materials

Snellen chart
measuring tape

Procedures

1. Fasten the Snellen chart on a suitable wall at eye level and make a mark on the floor exactly 20 feet from the chart.
2. Students should work in pairs and test each other's eyes. Measure one eye at a time and cover the eye not being tested. Test students who wear eyeglasses with and without their spectacles.

THE INGERSOLL EYE MODEL

The Ingersoll eye model allows us to study several fundamental optical properties of the human eye. Figure 13-3 shows the basic structure of the model and illustrates the various positions for insertion of lenses, the cornea, and three retinal positions. The kit also contains a variety of lenses.

FIGURE 13-3 Diagram of the Ingersoll (Cenco) eye model.

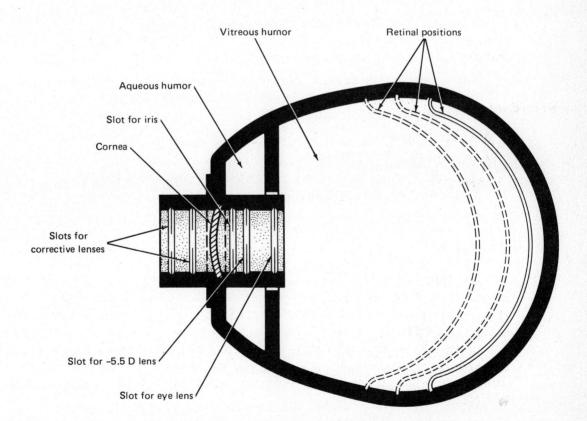

Materials

Ingersoll eye model with accessories (see Note)
1% eosin solution
rulers

Note: Available from Central Scientific Company, 2600 South Kostner Avenue, Chicago, IL 60623.

Procedures

"Accommodation"

1. Fill the tank of the eye model to within one inch of the top with dye solution (1% eosin in water). Place the "retina" in the central position (R).
2. Insert the +20D lens in position L of the model and adjust the distance of the lamp so that the image is sharply focused on the retina. Measure D_O and D_I (the distance from the lens to the object and to the image). Do these distances fit the basic lens equation?
3. What happens when the lamp is moved away from the lens? What change must you make in the lens to "accommodate" to this new distance.
4. Suppose that you moved the lamp *very* far away. What change would be needed?
5. Place the "cornea" in position C. How does this change the situation? Can you draw a lens diagram for this? (Recall that when two lenses are used in sequence, the image from the first lens becomes the object for the second.)
6. Does the presence of water in the tank (the "vitreous humor") alter light refraction?

Myopia and Hypermetropia

1. Place the retina in position R_m and insert lens +7D in position L. Is the image sharply focused in the retina? Where does it focus?
2. Which lens must you use to correct the condition of myopia? Can you draw a ray diagram to illustrate this?
3. Next place the retina in position R_h and insert lens +7D in position L. Is the image sharply focused? Where does it focus?
4. Which lens must you use to correct the condition of hypermetropia.

Astigmatism

1. Place the retina in position R. Examine the lens marked −5.50D by viewing a printed page through the lens and rotating it. What distortions occur? Place the lens in the position immediately behind the cornea.
2. Astigmatism in the human eye is ordinarily caused by slight curvature of the cornea. We have simulated this condition in the Ingersoll model with two separate lenses. What lenses are needed to correct this condition?
3. Can you arrange a lens system that is both astigmatic and myopic or astigmatic and hypermetropic? Can these be corrected?

Pupil Size

1. Set up the model with the +7D lens, the "cornea" in place, and the retina in position R. Move the lamp so that the image is somewhat out of focus.

Exercise 13]   Some Elements of Human Vision

2. Now place the "iris" with a 13 mm pupil in front of the cornea and observe the image. What happens?
3. Have a student in the class who wears eyeglasses for myopia remove his glasses and perform the following test. Ask him to look at a distant object (without his glasses), and then view the same object through an index card with a pinhole in it. What does he report? How does the pinhole improve visual acuity?

ACCOMMODATION IN THE HUMAN EYE

The capacity of the human eye to focus on very distant scenes and then to adjust rapidly and focus on close objects is termed accommodation. Very obviously the human eye contains only a single lens, and accommodation must be effected by altering the shape of this lens. Which muscles do this? (Recall that "accommodation" of the Ingersoll eye model necessitated changing the lens. Also, bear in mind that no photography enthusiast is satisfied with a single lens for his camera.)

With increasing age the human lens becomes much less elastic and cannot accommodate to close objects. The near point is defined as the shortest distance from the human eye at which an object is still in sharp focus.

Materials

pencils (or straight pins)
meter sticks
small clamps for attaching pencils to meter stick

Procedures

1. Attach two pencils (or pins) to the edge of a meter stick at distances of 25 and 100 cm from one end.
2. Hold the free end of the meter stick at the bridge of your nose and focus your gaze at the *near* pencil. How many pencils do you see behind the near pencil? Is it (are they) in focus?
3. Now focus on the *far* pencil. How many pencils can you see in front of it? Is it (are they) in focus?

Near Point of Accommodation
1. With your left hand hold a meter stick at the bridge of your nose directed forward. Take a pin in your right hand and hold it at arm's length. With one eye closed, slowly bring the pin along the edge of the meter stick toward your eye until it is no longer in sharp focus. The point at which the pin is still just focused is called the *near point*. Your partner will record your near point for each eye.
2. Compare the near points for each of your eyes with the average values below.

Age (years)	10	20	30	40	50	60	70
Near point (cm)	9	10	13	18	53	83	100

PROPERTIES OF THE PUPIL

Materials

> small flashlights
> matches

Procedures

1. Note the size of your partner's pupils. Observe what happens when a light is shone on your partner's eye. What happens in dim light? (If possible, repeat these observations later with a household pet, preferably a cat.)
2. Have your partner hold a book vertically in front of his face with the book binding along one side of his nose. Shine a fairly bright light on one eye, taking care that the other eye does not receive additional illumination.
3. Ask your partner to focus his sight on a small object (e.g., a match) held at arm's length. As the subject brings the object slowly toward his eyes, notice the size of his pupils. Describe what happens. Repeat the procedure, but this time observe the iris and pupil in profile view as the object is brought closer to the subject's eye. What happens?

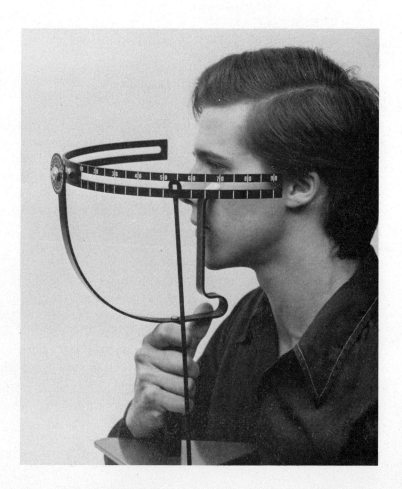

FIGURE 13-4 Perimeter for the measurement of peripheral vision.

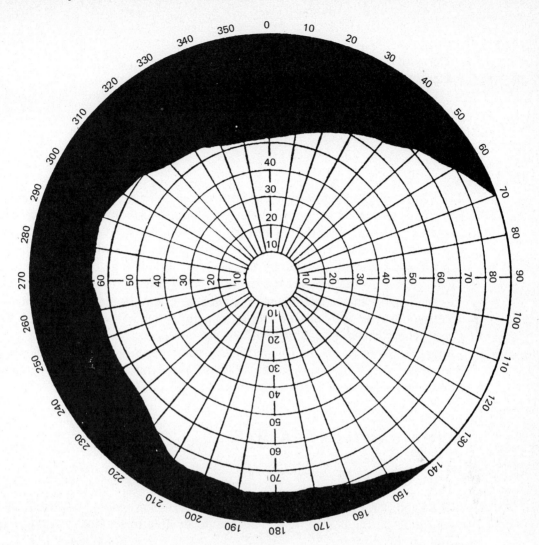

FIGURE 13-5 Perimetric chart showing the field of vision for the right eye while this eye looks directly forward and does not move. The numbers along the vertical and horizontal meridians are degrees of visual angle from the center of the fovea. What does the dark area represent?

PROPERTIES OF THE RETINA:
A. VISUAL FIELDS OF THE EYE

The field of vision encompasses all that one can see when the eye is held in a fixed position. Even when one's gaze is directed forward, considerable peripheral vision is possible. Figure 13-4 depicts the perimeter and Figure 13-5 illustrates the field of black and white vision for the right eye based on perimetric measurements. In this exercise we shall measure the visual fields of the eye for both black and white and color vision using a perimeter.

Materials

perimeter with rod and colored discs (see Note)
perimeter scoring charts (for right and left eyes)

Note: Available from Stoelting (see page 142).

Procedures

1. Have the subject assume a sitting position with the underside of his ocular orbit resting in the center of the perimeter. Ask him to look into the small mirror. The recorder sits in back of the device facing the subject.
2. Systematically test peripheral black and white vision by moving the pointer containing the white disc along the semicircular carriage of the perimeter from the outside inward. Record the angle at which the white disc can first be seen. Repeat this with the perimeter in a vertical orientation and at various angles. Score both eyes. Why are the scoring papers shaded in specific regions? Can you verify the existence of a blind spot?
3. Repeat the measurements for peripheral color vision using the blue, green, and red color discs. The subject must identify the particular color being tested. Randomize the order of color presentation in this test.

B. BLIND SPOT EXPERIMENT

Procedures

1. Close your left eye and hold this page about 20 inches in front of your face. Focus your gaze on the black circle. You should be able to see both the cross and the circle at this distance.

2. Gradually bring the paper closer to your face until the cross is no longer visible. Bring the paper even closer. Repeat the experiment for the other eye. What happens?

C. OPHTHALMOSCOPIC OBSERVATION OF THE RETINA

Under ordinary circumstances we are unable to view the retina of another person, because only a very limited amount of light passes through the pupillary aperture. Even if a bright light is brought near to the eye, the observer cannot see in because he blocks the light source as soon as he positions himself in front of the subject. The ophthalmoscope obviates this difficulty and is commonly used in clinical examination of the eye.

Exercise 13] Some Elements of Human Vision

Materials

ophthalmoscope

Procedures

1. The observer seats himself directly in front of the subject and looks at the right eye with his right eye, the left with his left. It is necessary to bring the viewing device extremely close to the subject's eye. **Exercise appropriate caution in performing this procedure.**
2. The ophthalmoscope is equipped with a series of small lenses (20D to -20D) that may be used to correct refractive errors in the subject's and/or observer's eyes. Consider that you are probably viewing an object (i.e., your partner's retina) that is closer to your eye than your near point.

STUDY QUESTIONS

1. What features of the eye determine visual acuity; that is, why cannot the normal human eye see the "E" on the Snellen chart at 2000 feet? Many birds have much greater visual acuity than man. How is a bird's eye adapted for this?

2. What is a phototopic curve and how is it measured? What does it indicate about the sensitivity of the human eye to various wavelengths of light? What is a scotopic curve?

3. Does binocular vision necessarily imply convergence? That is, do animals with binocular vision, such as arboreal primates, necessarily have the capacity to bring their lines of sight into convergence? Is convergence necessary for depth perception? How would you investigate these problems experimentally?

4. What processes go on in the eye when we shift our gaze from a far point to a near object? Does a similar process go on in the eyes of birds and other animals? What are some of the ways in which animals "accommodate"? (See von Heyningen [1975] and Walls [1942].)

5. Characterize each of the following defects or disorders that affect the human eye: cataract, ocular albinism, detached retina, night blindness, glaucoma, diabetes mellitus. What are the physiological consequences of each? What therapeutic measures can be used?

6. What regions of the brain are involved in vision? What is the condition known as ideational agnosia and what is its organic basis? What does the condition show about the relation of perception to cognition?

7. Figure 13-6 illustrates the basic optics of a simple convex lens. The ray diagram shows that light rays from the object O are caused to bend and converge by the lens. The rays cross to produce an *inverted, real, diminished* image (I). Point F is the focal length of the lens in centimeters. D_o is the distance from the lens to the

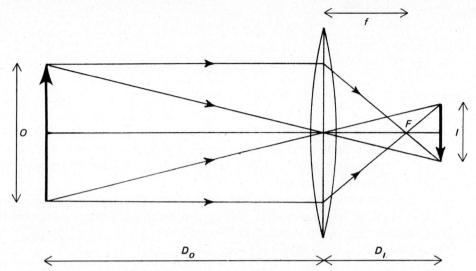

FIGURE 13-6 Ray diagram of a convex (positive) lens showing F, the principal focus; f, the focal length; D_O, the distance to the object; D_I, the distance to the image. O is the object, I is the image.

object; D_I is the distance from the lens to the image. Two well-known equations, derived in any elementary physics text, show that

$$\frac{I}{O} = \frac{D_I}{D_O} \quad \text{and} \quad \frac{1}{D_O} + \frac{1}{D_I} = \frac{1}{F}$$

Use these relationships to solve the following problems.

a. Determine by calculation the position of the image formed by a lens of focal length +15 cm as the object distance is made, in turn, 20, 30, and 60 cm. Calculate the distance between the object and image for each case. Draw ray diagrams to scale to illustrate each case. What would happen if the object was moved very far away (i.e., so that D_O was essentially infinite and $1/D_O$ reduced to 0)? Draw a ray diagram to illustrate this.

b. A student with a convex lens of 18 cm focal length wanted to produce an image equal in size to the object. Find the appropriate object distance (D_O) and image distance (D_I). Draw a ray diagram to illustrate this problem. (*Hint*: Consider the two equations above. What happens if $I = 0$?)

c. Draw a diagram of the human eye for a person who is hypermetropic. What kind of lens is used to correct this condition? Illustrate by a ray diagram including both the lens of the eye and the corrective lens. Make a comparable drawing for a myopic individual.

REFERENCES

Brown, K. T. 1968. The electroretinogram: Its components and their origins. *Vis. Res., 8*:633.

Chapanis, A. 1975. Visual pigments and color blindness. *Sci. Amer., 232*(3):64.

Cornsweet, J. N. 1970. *Visual Perception.* Academic Press, New York.

Davson, H. (ed.), 1962. *The Eye* (5 vols). Academic Press. New York. (Contains many excellent articles on all aspects of vision.)

Daw, N. W. 1973. Neurophysiology of color vision. *Physiol. Rev., 53*:571.

Fox, R., et al. 1976. Falcon visual acuity. *Science, 192*:263.

Padgham, C. A., and J. E. Saunders. 1975. *The Perception of Light and Color.* Academic Press, New York.

Pettigrew, J. D. 1972. The neurophysiology of binocular vision. *Sci. Amer., 227*(2):84.

Senders, V. L. 1948. The physiological basis of visual acuity. *Psychol. Bull., 45*:465.

Toates, F. W. 1972. Accommodation function of the human eye. *Physiol. Rev., 52*:828.

von Heyningen, R. 1975. What happens to the human lens in cataract. *Sci. Amer., 233*(6):70.

Walls, G. L. 1942. *The Vertebrate Eye.* Cranbrook Institute of Science, Bloomfield Hills, MI.

EXERCISE 13

Name _____

Laboratory Section _____ Date _____

RESULTS AND CONCLUSIONS

Binocular Vision

Explain your observations with the bead and string. How does this simple apparatus illustrate convergence of the two visual axes? Under what circumstances do you see two strings? Explain this phenomenon.

Depth Perception

Average deviation with both eyes open_____

Average deviation with left eye_____

Average deviation with right eye_____

What do you conclude from these measurements?

Eye Preference

Which is your dominant eye? Does your eyedness correlate with ?dedness? Collect and analyze the class data.

153

Exercise 13] Results and Conclusions

Visual Acuity

If you wear eyeglasses, enter the uncorrected visual acuity for each eye.

Left eye: 20/_____

Right eye: 20/_____

Ingersoll Eye Model

Tabulate and thoroughly discuss the experiments and observations you have made with the Ingersoll model.

Accommodation in the Eye

Discuss accommodation in the human eye. What change is occurring in the lens to permit you to focus at different distances? Draw a simple lens diagram, including the two objects, with the eye accommodated to either the near or far pencil (finger). Where do the images of both pencils fall?

Near point (left eye) _____ cm

Near point (right eye) _____ cm

Compare your near points with the data given on page 145. Is there a difference in the near points for your eyes? Is this related to eye dominance?

Properties of the Pupil

Explain your observations on pupillary responses. What is the pupillary reflex? What portion of the nervous system is responsible for this reflexive action?

Properties of the Retina

Compare the visual fields for the three colors and for black and white vision. What conclusions can you draw from these observations? What is the anatomical basis of the blind spot?

EXERCISE 14

Hearing and Balance

OBJECTIVES In the first series of experiments in this laboratory exercise we shall (1) measure auditory acuity, both qualitatively and quantitatively; (2) make a simple assessment of sound localization; (3) investigate direct bone conduction in the ear (Rinne Test); and (4) study auditory fatigue. In the second part we shall study the sense of balance and its relation to other senses.

Through the sense of hearing we are placed into direct, intimate contact with the surrounding world. Musical, vocal, and other sonic impressions flood us constantly. We possess a wealth of empirical data concerning the reception and transmission of sound in the human ear. We also have information about the anatomical and neurophysiological basis of these processes; but we do not have a satisfactory, clearly defined understanding of "how we hear." Several of the references at the end of this exercise treat the various theoretical proposals that have been offered to explain the process of hearing and present the current views and problems at the forefront of a very complex area of biological research.

Figure 14-1 is a drawing of the human head showing the auditory apparatus. Study the illustration and identify the tympanic membrane, the three auditory ossicles, the cochlea, semicircular canals, Eustachian tube, and the pharynx.

The normal human ear is sensitive to sonic frequencies ranging from about 20 to 20,000 Hertz (Hz; 1 Hz = 1 cycle/second), although the range varies with age and other individual factors. Several mammalian species, including bats, dolphins and whales, and many rodents, can detect much higher frequencies. Sensitivity of the human ear varies with the sonic frequency. Hearing in man is most sensitive in the range 500-4000 Hz. Figure 14-2, compiled from U. S. Public Health Service data, illustrates auditory thresholds for human hearing and shows the relative proportion of individuals at or below each threshold level. The 1% level indicated in the graph constitutes the "ideal human ear."

AUDITORY ACUITY

The determination of auditory acuity is ordinarily a complex procedure, but a very simple test can be performed with a ticking watch.

158 Section IV] Nerve and Sensory Processes

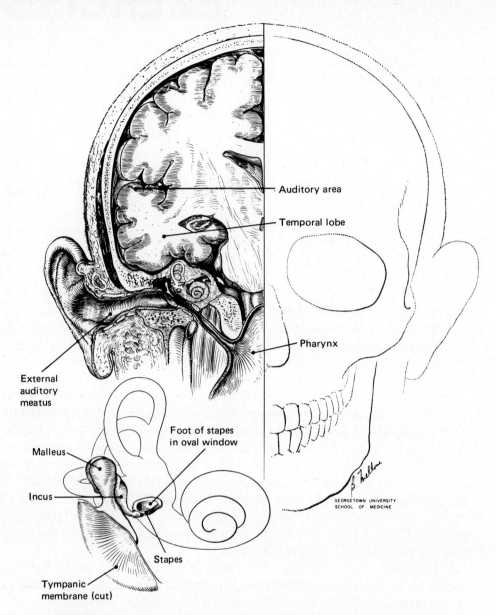

FIGURE 14-1 The human auditory system. Identify the various components as indicated in the text. [*Drawn by B. Melloni and reproduced with permission from* The Human Body: Its Structure and Physiology *by S. Grollman, 3rd ed., 1974, Macmillan Publishing Co., Inc., New York.*]

Materials

      watch or alarm clock with a loud tick
      meter stick
      megaphone or other large conical object
      wax earplugs (such as dispensed by pharmacists to facilitate sleep) or sterile cotton

Exercise 14] Hearing and Balance

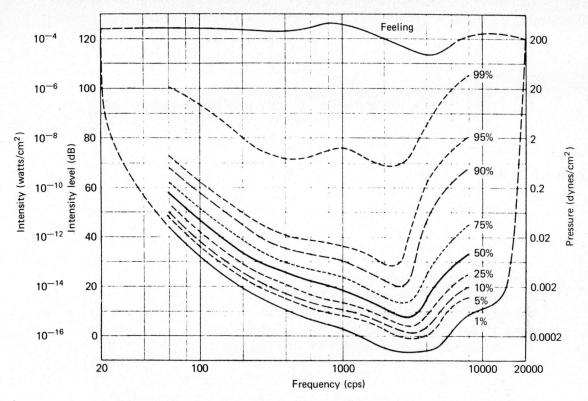

FIGURE 14-2 Auditory thresholds of the human ear as a function of sonic frequency. The various curves indicate the percentages of a typical American population who can hear at or below the levels shown in the figure.

Procedures

1. Work in pairs for this exercise. The background noise in the room should be minimal.
2. Plug one ear and have your partner hold the ticking watch in front of the open ear. As the watch is moved away from your ear, listen carefully and indicate when you can no longer hear the ticking. Your lab partner should measure and record this distance. Repeat the measurement several times and average the results. Test the other ear.
3. Can you improve your auditory acuity, as measured by this test, by listening through a megaphone? Does cupping your hand over your ear assist?

SOUND LOCALIZATION

Materials

clicking device (such as a dimestore "cricket")
blindfold

Procedures

1. Work in pairs. The subject should be blindfolded and seated in an open area.
2. As the "cricket" is clicked at various positions, the subject should point to where he thinks the sound is originating.
3. Repeat the test with one of the subject's ears closed with an ear plug. Is sound localization affected?

DIRECT BONE CONDUCTION AND THE RINNE TEST

The conduction of sound through the middle ear is facilitated by the concerted action of the three smallest bones in the human body, the auditory ossicles. The outermost bone, the malleus (hammer), is secured to the inner aspect of the eardrum itself. The innermost bone, the stapes (stirrup), is positioned with its foot plate in the oval window of the cochlea. The incus (anvil) connects the two. These three bones respond to external impressions coming to the eardrum from the surrounding world and transmit the delicate vibrations to the fluid within the cochlea. This aspect of the hearing process is ordinarily referred to as the conductive phase.

Sound can also be transmitted to the fluid of the cochlea by direct conduction through the bones of the skull, i.e., by a process in which no vibrations pass through the auditory ossicles. This kind of direct bone conduction provides an alternative pathway for sound transmission in individuals with middle-ear defects. To some extent you hear the sound of your own voice through the mechanism of direct bone conduction.

The Rinne test, originally designed by the German otologist, Dr. A. Rinne, differentiates between conductive and sensorineural hearing impairments, and is an important aspect of auditory diagnosis. A vibrating tuning fork is held against the mastoid process of the temporal bone (the knob at the base of the ear) until the subject can no longer hear the tone, at which time the investigator quickly places the fork in front of the ear. Someone with normal hearing can still hear the tone for a short time by direct air conduction and is said to be Rinne positive. Individuals with middle-ear defects usually are Rinne negative; that is, they hear better with temporal bone conduction than with conduction through the middle ear. Individuals with sensorineural defects (defects of the sensory mechanism in the cochlea itself, defects in the cranial nerve VIII, or central nervous system damage) are Rinne positive; that is, they can hear much better on a relative basis through air conduction than through bone. The auditory thresholds for Rinne positive, however, are much higher.

Materials

tuning forks (128, 256, 512, and 1024 Hz)
wax earplugs or sterile cotton

Procedures

1. Direct bone conduction of sound can be demonstrated by placing the handle of an inaudible vibrating tuning fork in the middle of the forehead or in between the

front teeth. The ears should be stopped with wax earplugs or with the little fingers.
2. Students should work in pairs for this test; one as subject, the second as investigator. Plug the ear that is not being tested. The investigator lightly strikes the tuning fork with a soft mallet (or his palm), and presses it to the mastoid process of the subject. As soon as the subject reports he can no longer detect the sound, the investigator positions the fork in front of the external ear. If the subject can hear the sound at this point, he is Rinne positive. Both ears are to be evaluated with several tuning forks of different frequencies.
3. The test should also be conducted in the reverse manner; that is, air conduction followed by bone conduction, to verify the conclusions.
4. Middle-ear defects can be simulated by plugging the auditory canal with earplugs. Can you demonstrate a Rinne negative test in this manner?

AUDITORY ADAPTATION

The human senses of taste, smell, perception of warmth, and other senses as well, show fatigue or adaptation. We will investigate this phenomenon for hearing by carrying out the procedures described below.

Materials

stethoscope
tuning forks

Procedures

1. Students should work in pairs; one as subject, the second as investigator.
2. The subject places a stethoscope in his ears and seats himself in front of the investigator. The investigator strikes the tuning fork lightly with a soft mallet or with the palm of his hand and holds it against the bell of the stethoscope so that the sounds received in each ear are approximately equal.
3. After a brief interval, the investigator then firmly pinches one tube of the stethoscope and repeats procedure 2. When the sound is reported by the subject to have diminished, the investigator releases the stethoscope tube, strikes the tuning fork again, and applies it to the stethoscope bell. What does the subject report?
4. Does auditory adaptation to one frequency generalize to other frequencies? How could you test this?

AUDIOMETRY

The quantitative clinical evaluation of hearing capacity is ordinarily performed with a device known as the pure tone audiometer. This instrument produces tones of the following frequencies: 125, 250, 500, 750, 1000, 1500, 2000, 3000, 4000, 6000, and

8000 Hz. In addition, the intensity can be regulated by means of an alternator so that auditory thresholds at each frequency can be determined. In practice, each ear is tested at each frequency until the threshold is found; these values are recorded in graphic form as the audiogram.

It is especially important to note that the audiometer is precalibrated for each sonic frequency to the threshold of the "average normal ear" as a reference. A subject just able to perceive +5 decibels (dB) at frequency 2000 Hz can hear a tone 5 dB louder than one perceived by the "average normal ear" at this frequency. Individuals with acute hearing may actually hear better than those with the average ear. In other words, the audiometer does not measure absolute auditory thresholds; it simply evaluates the subject relative to what the average human ear can perceive. See Figure 14-2 for a graphic representation of the innate sensitivity of the human ear at various frequencies.

Figure 14-3 illustrates a series of audiograms for an individual who has been exposed to high intensity industrial noise for a period of twenty years. Both ears were affected similarly, although only one is represented in the diagram.

Audiometers are usually equipped to measure both air and bone conduction. This allows one to perform Rinne's test in a quantitative fashion and to compare relative hearing capacities by the two modes of conduction.

FIGURE 14-3 Audiograms showing typical hearing loss after varying exposures to high intensity industrial noise.

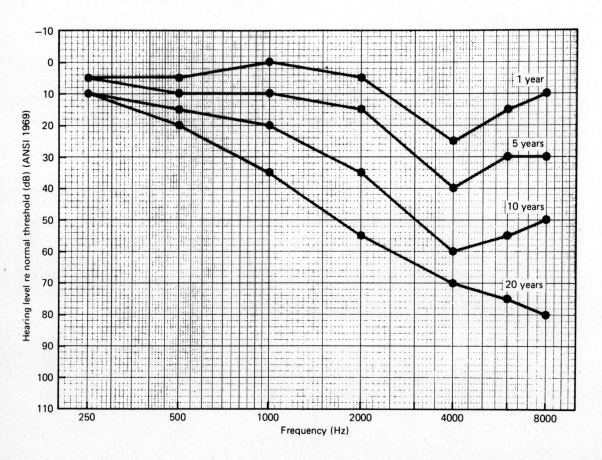

Exercise 14]   Hearing and Balance

Materials

pure tone audiometer (see Note)
quiet room (soundproof room is preferable but not essential)

Note: Available from Beltone Electronics Corporation, 4201 West Victoria, Chicago, IL 60646; Maico Hearing Instruments, 7375 Bush Lake Rd., Minneapolis, MN 55435; and other manufacturers.

Procedures

1. The audiometer should warm up 10 minutes prior to use. The earphones should be plugged into the audiometer.
2. Students may work in pairs. The investigator places the earphones snugly over the subject's ears and sets the frequency control to 1000 Hz, the intensity control to 40 dB. A subject with normal hearing will be able to perceive the tone as the intensity is turned from 40 down to +5 or 0. This procedure is simply a check on the instrument and the earphones. Test both earphones.
3. Begin the test at 1000 Hz, as this point is near the point of greatest sensitivity of the human ear. The tone will be inaudible except when the interrupter switch on the audiometer is depressed. Start with the intensity control at its minimum value (0 dB), depress the interrupter switch, and increase the level until the subject signals that he can hear the sound. Go about 10-20 dB higher so that the subject can hear the pitch of the tone well. Release the interrupter switch.
4. Decrease the intensity in 5 dB intervals below this value, but only depress the interrupter switch for about 1 second for each trial. Record the level at which the sound is last heard. Test both ears.
5. Proceed to frequencies of 500 and 250 Hz. Next, evaluate hearing at the higher frequencies (1500, 2000, 3000, 4000, 6000, 8000 Hz). Construct the audiogram.
6. If time permits, repeat the procedure using the bone-conduction device supplied with the audiometer.

BALANCE (EQUILIBRIUM)

We all are aware that in sitting, reclining, or standing erect we have a very particular sense of being in a state of balance. Without this sense we would be unable to maintain a fully alert conscious life, as anyone knows who has experienced disequilibrium or vertigo. The sense of balance or equilibrium is dependent upon the proper functioning of an organ of the inner ear, the semicircular canals, but is also associated with sight, hearing, and the proprioceptive sense (the awareness we have of our body movements and of where our limbs are located in space).

We shall now perform two simple experiments to investigate the sense of balance.

Materials

swivel chairs
stopwatches

164 Section IV] Nerve and Sensory Processes

Procedures

1. Select one student as a volunteer and seat him on a swivel chair in the center of an open area. Position several students around the chair to prevent the volunteer from falling during certain phases of these investigations.
2. One student should spin the chair for 10–12 turns at a moderately slow rate (1 turn per 2 or 3 seconds). The subject should have his eyes closed throughout the procedure, and should keep his head in a fully upright position.
3. After a specified number of turns has been completed, chair rotation should be stopped. Record the time it takes for the subject to report the cessation of movement. In this and all subsequent procedures, be certain to record all subjective sensations the volunteer experiences.
4. Allow the subject to recover equilibrium and repeat the experiment with the subject's eyes open. Are there any differences?
5. Repeat the procedure as in step 3 above, but ask the subject to open his eyes as soon as the chair has been stopped. Observe his eye movements.
6. Repeat step 3, but with the subject's head oriented in different planes (that is, with his chin resting on his chest, with his head alternately on his right and left shoulder, with his neck bent backward). In each of these cases, ask the volunteer to stand immediately after the chair stops, and notice carefully in which direction he tends to fall. (The other students should prevent the volunteer from falling.)
7. What conclusions can you draw from these simple observations?

Materials

blindfolds
swivel chairs

Procedures

Students will work in groups of three for this experiment.
1. Tape a sheet of paper to the floor of the laboratory room and draw the outline of the subject's foot on the paper.
2. The subject should stand on the sheet with one foot and touch the other foot to his knee in any way he chooses.
3. The test will measure the length of time the subject can remain standing in this position (with one foot within the drawn outline and the other foot touching the knee) to a limit of 90 seconds. One student will act as timer; the second will observe the movements of the subject and call out the moment when the trial is terminated.
4. With the procedure outlined above, test the ability of the subject to balance himself under four separate conditions.
   a. With eyes open and ears unplugged.
   b. With eyes blindfolded and ears unplugged.
   c. With eyes open and ears plugged.
   d. With eyes blindfolded and ears plugged.
5. It is probably wise to rotate the subject, timer, and referee after each trial to minimize the effects of fatigue.

6. Collect the class data and analyze the results. What conclusions can you draw concerning the interaction of hearing and sight with balance?

STUDY QUESTIONS

1. Why does a tape recording of your own voice (even a good one) seem unusual to you? Why do many people with middle-ear damage tend to speak more softly? What kind of hearing damage might be found in a person who speaks in an unusually loud voice?

2. How are sound waves transmitted through the cochlea? What are the true auditory receptors and where are they located? What constitutes a "normal stimulus" for these receptors? Where are high-pitched sounds received? Low-pitched sounds? What is the actual experimental evidence in support of your contentions?

3. What is the acoustical reflex and how does it compare with the pupillary response of the eye?

4. What damage is associated with chronic exposure to loud noise or rock music? How does this show up in the audiogram? Why isn't the sensitivity to all frequencies affected equally?

5. Why is hearing temporarily impaired when the Eustachian tube is blocked during a head cold?

6. Why don't figure skaters and ballet dancers get dizzy?

7. Where is the auditory center of the brain? Where is the equilibrium center of the brain? Which nerve or nerves serve each of these senses? What is the organic basis of auditory agnosia and what does this indicate about perception and cognition?

REFERENCES

Brown, J. L. 1961. Orientation to the vertical during water immersion. *Aerospace Med., 32*:209.

Davis, H., and S. R. Silverman (eds.). 1973. *Hearing and Deafness,* 3rd ed. Holt, Rinehart and Winston, New York.

Eldredge, D. H., and J. D. Miller. 1971. The physiology of hearing. *Ann. Rev. Physiol., 33*:281.

Fletcher, H. 1953. *Speech and Hearing in Communication,* 2nd ed. D. Van Nostrand, New York.

Flock, A., et al. 1962. Morphological basis of directional sensitivity of the outer hair cells in the organ of Corti. *J. Acoust. Soc. Am., 34*:1351.

Goldberg, J. M., and C. Fernandez. 1975. Vestibular mechanisms. *Ann. Rev. Physiol., 37*:129.

Gulick, W. L. 1971. *Hearing.* Oxford University Press, New York. (An excellent, basic review of our present understanding.)

Harris, J. D. 1974. *Anatomy and Physiology of the Peripheral Hearing Mechanism.* Bobbs-Merrill, Indianapolis.

Jerger, J. (ed.). 1963. *Modern Development in Audiology.* Academic Press, New York.

Moller, A. R., and P. Bastan (eds.). 1973. *Basic Mechanisms in Hearing.* Academic Press, New York. (Symposium papers giving a broad treatment of audition.)

Newby, H. A. 1972. *Audiology,* 3rd ed. Prentice-Hall, Englewood Cliffs, NJ.

Rosenzweig, M. R. 1961. Auditory localization. *Sci. Amer., 205*(4):132.

Ward, W. D. 1968. Susceptibility to auditory fatigue. In W. D. Neff (ed.), *Contributions to Sensory Physiology,* Vol. 3. Academic Press, New York.

EXERCISE 14

Name _____

Laboratory Section _____ Date _____

RESULTS AND CONCLUSIONS

Auditory Acuity

What is your auditory acuity, as measured with a ticking watch, in each ear? What procedures improve your acuity?

Sound Localization

Describe your results. Are there specific directions in which sound localization is more exact? Is the localization of sound affected by plugging one ear?

The Rinne Test

Are you Rinne positive or negative in each ear? If you are Rinne positive, have you been able to simulate the Rinne negative condition?

Auditory Adaptation

Describe your results from this test. Is auditory adaptation specific for the original frequency or can it be generalized? How does auditory adaptation compare with similar phenomena for other senses?

Audiometry

Prepare your own audiogram for each ear by plotting the data you have obtained. Plot sound frequency in Hz on the abscissa, hearing level in dB on the ordinate (with −10 dB at the top of the scale and 10 dB increments in descending order). Interpret your audiogram. Are there any indications of hearing impairment? (Newby [1972, Chapter 5] presents a good introductory discussion of normal and abnormal audiograms and their interpretation.) How do the thresholds for bone conduction compare with those determined for air conduction?

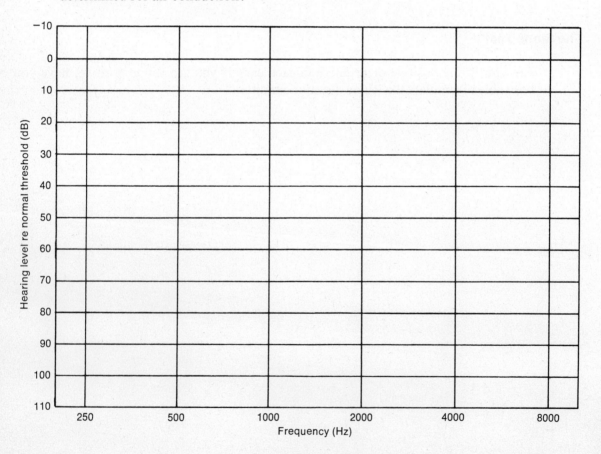

Balance (Equilibrium)

Describe and interpret the results of the procedures dealing with equilibrium. What other senses are involved in maintaining balance?

SECTION V

Digestion

EXERCISE 15

Digestion

OBJECTIVES In this exercise we shall investigate several of the chemical processes taking part in normal digestion, such as (1) salivary digestion of starch; (2) pancreatic digestion of fats and the role of bile salts in this process; (3) the action of rennin; and (4) the digestion of protein. An anatomical illustration of selected regions of the gastrointestinal tract is provided to indicate where these digestive processes occur.

A constant supply of nourishing substances is an essential requirement for all living organisms. Through digestive functions, materials taken into the body are degraded, dissimilated, and liquefied. In the process food substances lose all semblance of their former state and are reduced to a very fine consistency. In this way digestion and assimilation help to prepare the body's own substance.

Digestion begins in the mouth. It is here that food is mechanically broken down into smaller particles and ensalivated prior to swallowing. The digestive enzyme ptyalin (salivary amylase) acts to decompose starches in the mouth. Maltase, an enzyme catalyzing the breakdown of the disaccharide maltose, is also a component of saliva. Assimilation also begins in the mouth, where the mucus membranes lining the oral cavity absorb sugars and transmit them directly to the blood.

After food has been thoroughly chewed and lubricated by the salivary secretions, it enters the pharynx and the process of swallowing is initiated. Involuntary peristaltic contractions take over as the food is passed along the esophagus to the stomach. Although the stomach is closed off at the anterior end by the cardiac sphincter, this valve opens as food materials approach and food is allowed to enter the stomach.

The first region of the stomach, the fundus, serves mainly for storage. Gastric digestion proper takes place in the pylorus, a highly active region in which vigorous churning mixes the food with gastric secretions. The stomach contents are prevented from leaving the organ by the pyloric sphincter, and food may remain within the stomach for several hours. In time the sphincter opens and allows portions of chyme to enter the small intestine. It is probably accurate to state that one of the most important functions of the stomach is storage. Gastrectomized individuals can lead fairly normal lives, but they must eat smaller, more frequent meals.

Gastric juice is produced by many tiny glands lining the stomach wall and contains several digestive elements. Pepsinogen, an inactive form of the enzyme pepsin, is secreted into the stomach where it is converted to an active form by the highly acidic stomach contents. This enzyme helps to degrade protein. Hydrochloric acid at pH 1-2 is secreted by cells in the stomach lining and forms a chief component of the gastric fluid. Rennin, an enzyme that causes milk to curdle, is present in the stomach, especially in infants, as is lipase, a fat-digesting enzyme that is a smaller component of gastric juice.

The small intestine is an immense organ, often 25 feet in length. There are three major subdivisions of the small intestine:

(1) The duodenum, about 1 foot in length.
(2) The jejunum, approximately 5 feet in length.
(3) The ileum, the major portion of the intestine.

The stomach contents first enter the duodenum; it is here that ulceration often occurs, possibly in part because of the highly acidic nature of the gastric fluids. Within the duodenum, the semifluid mass of partially digested food mixes with secretions from two highly important organs, the pancreas and the gall bladder. The alkaline nature of the pancreatic juice helps to keep the normal duodenal pH at a value of about 8. Can you identify and label the pertinent structures in Figure 15-1?

The flow of pancreatic juice is regulated by the hormone secretin, the production of which is induced by the presence of acid chyme within the duodenum. Pancreatic juice contains many digestive enzymes, including lipase, amylase, trypsin, and chymotrypsin. The latter two enzymes, both of which catalyze the degradation of protein, are secreted as inactive forms and become activated in the duodenum.

Bile is a digestive fluid that enters the duodenum from the gall bladder by way of the common bile duct. Although this secretion contains no digestive enzymes, it plays a highly important role in the emulsification of fatty substances. Emulsified fats are more easily acted upon by pancreatic lipase. The flow of materials from the gall bladder is regulated by a sphincter which is under hormonal control. The presence of fats in the duodenum induces the production of cholecystokinin, a hormone that relaxes the cystic sphincter and induces contraction of the muscular walls of the gall bladder.

As the semifluid mass continues through the jejunum, it mixes with secretions from the wall of the intestine itself. The intestinal secretions, known as *succus entericus,* contain mucus, which assists in the digestive flow, and a variety of catabolic enzymes, such as peptidases, disaccharidases, and lipase. These enzymes bring to completion the processes previously started in the gastrointestinal tract.

Assimilation is a chief function of the small intestine. The countless folds in the intestinal lining afford a vast surface area through which absorption can occur. Finger-like processes, called villi, each of which contains a tiny capillary loop, are specialized for the assimilation of digested food materials. Both simple diffusion and active transport are involved in uptake into the blood. In addition, each villus contains a tiny lymphatic vessel called a lacteal. Amino acids and sugars pass into the capillaries of the villi; fatty acids, glycerol, and glycerides are absorbed into the lacteals.

At the juncture of the small intestine with the large intestine, the ileocaecal valve prevents material from backing up into the small intestine. Within the large intestine, which is subdivided into the ascending, transverse, and descending colons, there exist teeming numbers of microorganisms. This intestinal flora further degrades undigested materials, and helps to provide essential vitamins (such as vitamin K and some B vitamins) needed by the human organism. The absorption of large amounts of water also occurs in the large intestine. The residue remaining from this process passes into the rectum and is eliminated.

Exercise 15] Digestion

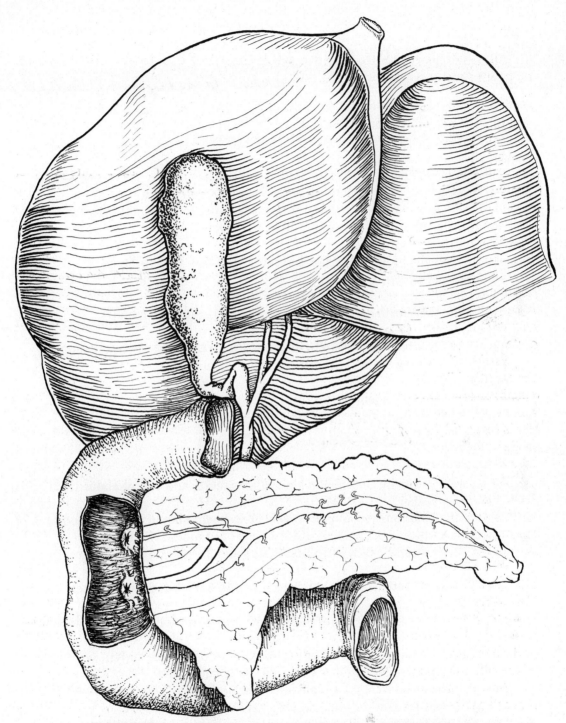

FIGURE 15-1 Inferio-posterior aspect of the liver and gall bladder. Identify and label the following structures: duodenum, liver, gall bladder, pancreas, hepatic duct, cystic duct, common bile duct, pancreatic duct, accessory pancreatic duct, and major and minor duodenal papillae.

In this exercise we shall investigate several of the chemical processes involved in digestion.

SALIVARY DIGESTION OF STARCH

Materials

make 6 setups 3/students per group

test tubes (6 per ~~two~~ 3 students)
test tube racks (1 per ~~two~~ 3 students)
marking pencils
~~37°C water baths (1 per four to eight students)~~ *use 37°C incubator in micro)*
~~Pasteur~~ pipets (4 per two students) *medicine droppers.*
boiling water baths equipped with test tube racks (1 per four to eight students)
Bunsen burners and tripods (1 per four to eight students)
spot plates (1 per ~~two~~ 3 students)
beakers (or buckets) of ice (1 or 2 per laboratory section) *(maybe you should get this, this afterno)*
100 ml beakers (1 per two students)
Benedict's qualitative reagent (50 ml per two students)
Lugol's iodine solution in dropping bottles (1 bottle per two students)
parafilm squares *(1 per person*
1% solution of soluble starch (heat to dissolve starch) (20 ml per two students)
0.1 N HCl in dropping bottles (1 bottle per two students)
distilled water
~~5 ml pipets (3 per two students)~~

graduates (sm)
beakers (sm)

Procedures

1. Induce salivation by chewing on a parafilm square, and collect about 15–20 ml of saliva in a beaker.
2. All of the solutions to be used in this experiment should be preheated to 37°C except as indicated below. Prepare a series of test tubes as follows:
   a. To tube #1 add 2.0 ml of soluble starch and 2.0 ml of distilled water. Incubate at 37°C.
   b. To tube #2 add 2.0 ml of soluble saliva and 2.0 ml of distilled water. Incubate at 37°C.
   c. To tube #3 add 2.0 ml of saliva and 2.0 ml of soluble starch. Incubate at 37°C.
   d. To tube #4 add 2.0 ml of saliva and 2.0 ml of soluble starch, *both of which solutions have been precooled to 0°C.* Incubate at 0°C.
   e. To tube #5 add 2.0 ml of saliva to which 0.1 N HCl has been added (5 drops acid/ml saliva) and 2 ml of soluble starch. Incubate at 37°C.
   f. To tube #6 add 2.0 ml of saliva, *which has previously been heated in a boiling water bath for several minutes and cooled to 37°C,* and 2.0 ml of soluble starch. Incubate at 37°C.
3. Let the six tubes incubate for 30 minutes. At the end of the 30 minute period, transfer each tube to a boiling water bath for 2 minutes. If you have started the reactions at different times, be certain to remove each tube after exactly 30 minutes incubation.

Exercise 15] Digestion 177

4. Place 1 drop from each tube in a separate compartment of a spot plate. Add 1 drop of Lugol's iodine to test for starch. A deep blue-black color indicates the presence of starch. Dextrins, intermediate products of starch degradation, give a red-purple color. The yellow color of iodine indicates complete hydrolysis of the starch. Record the results for each of the six tubes.
5. To each of the six test tubes add 2 ml of Benedict's qualitative reagent and place the tubes in a boiling water bath for 2 minutes. Remove the tubes and record the results.

PANCREATIC DIGESTION OF FATS

Litmus, the commonly used pH indicator, is blue in basic solution, red in acid. It can be used to study the enzymatic breakdown of fats because fatty acids are produced from the hydrolysis of fats.

Materials

heavy cream (one half pint per laboratory section)
litmus powder (finely ground with mortar and pestle)
1% pancreatin dissolved in 0.2% Na_2CO_3* (50 ml per two students)
test tubes (6 per two students)
test tube racks (1 per two students)
37°C water baths
distilled water
5 ml pipets (3 per two students)
bile salts
wax marking pencils

Procedures

1. Add enough litmus powder to a portion of heavy cream to produce a distinct bluish color.
2. Preincubate the litmus cream and pancreatin at 37°C. Prepare a series of test tubes as follows:
   a. To tube #1 add 2.0 ml of litmus cream and 2.0 ml of pancreatin. Incubate at 37°C.
   b. To tube #2 add 2.0 ml of litmus cream and 2.0 ml of distilled water. Incubate at 37°C.
   c. To tube #3 add 2.0 ml of litmus cream, 2.0 ml of pancreatin, and a small quantity of bile salts (the amount on the tip of a small spatula). Incubate at 37°C.
   d. To tube #4 add 2.0 ml of litmus cream, 2.0 ml of distilled water, and a small quantity of bile salts. Incubate at 37°C.
3. After 30 minutes incubation, inspect the tubes and record the results. Be certain also to smell each tube.

*If the pancreatin does not dissolve completely, the undissolved residue can be allowed to settle and the clear supernatant used. Centriguation can also be used to clarify the preparation.

THE ACTION OF BILE SALTS

Materials

test tubes (2 per two students)
test tube racks (1 per two students)
5 ml pipets (2 per two students)
distilled water
bile salts
vegetable oil (10 ml per two students)
wax marking pencils

Procedures

1. Prepare two test tubes as follows:
   a. To tube #1 add 3.0 ml of vegetable oil and 3.0 ml of distilled water.
   b. To tube #2 add 3.0 ml of vegetable oil, 3.0 ml of distilled water, and a tiny amount of bile salts.
2. Thoroughly mix both tubes for the same length of time and with identical agitation. Carefully observe the tubes for several minutes and note the results.

THE ACTION OF RENNIN

Materials

test tubes (10 per two students)
test tube racks (1 per two students)
wax marking pencils
5 ml pipets (2 per two students)
rennin tablets (see Note)
37°C water baths (1 per four to eight students)
boiling water bath (1 per four to students)
0.01 M sodium oxalate (2 ml per two students)
0.01 M sodium chloride (2 ml per two students)
0.1 M calcium chloride (2 ml per two students)
distilled water
buckets of ice (2 per laboratory section)

Note: Rennin tablets may be purchased as Junket (rennet) in pharmacies, health food stores, and some supermarkets. Individual tablets usually contain calcium phosphate buffer as well as a starch filler. Powdered Junket can also be used, although it is somewhat less convenient.

Exercise 15] Digestion

Procedures

1. Add a single rennin tablet to 10 ml of distilled water. The materials will probably not dissolve completely, but you should agitate the tube to assist dissolution.
2. All reagents, including the milk, should be prewarmed to 37°C except as indicated. Prepare a series of tubes as follows:
   a. To tube #1 add 3.0 ml of milk. Incubate at 37°C.
   b. To tube #2 add 3.0 ml of milk and 3 drops of rennin. Incubate at 37°C.
   c. To tube #3 add 3.0 ml of milk, 3 drops of rennin, and 3 drops of sodium oxalate solution. Incubate at 37°C.
   d. To tube #4 add 3.0 ml of milk, 3 drops of rennin, and 3 drops of sodium chloride solution. Incubate at 37°C.
   e. To tube #5 add 3.0 ml of milk, and 3 drops of rennin, *both precooled to 0°C.* Incubate at 0°C.
3. Allow the reactions to proceed for 10 minutes. At the end of this time inspect each tube for coagulation by tilting the tube. Tabulate your results. If tube #3 has not coagulated, add 3 drops of calcium chloride and incubate for a further 10 minutes. If tube #5 has not coagulated, incubate it at 37°C for 10 minutes.

DIGESTION OF PROTEIN

Materials

test tubes (10 per two students)
test tube racks (1 per two students)
wax marking pencils
5 ml pipets (4 per two students)
1 ml pipets (3 per two students)
37°C waterbaths (1 per four to eight students)
1% pancreatin dissolved in 0.2% Na_2CO_3 (10 ml per two students)
5% or saturated pepsin solution (10 ml per two students)
0.2% Na_2CO_3 (2 ml per two students)
0.4% HCl (20 ml per two students)
distilled water
carmine fibrin*

Procedures

1. Set up a series of test tubes as follows:
   a. To tube #1 add a small pea-sized chunk of carmine fibrin, 3 ml of pepsin, and 3 ml of 0.4% HCl.
   b. To tube #2 add carmine fibrin, 3 ml of pepsin, and 3 ml of distilled water.
   c. To tube #3 add carmine fibrin, 3 ml of distilled water, and 3 ml of 0.4% HCl.
   d. To tube #4 add carmine fibrin, 1 ml of pancreatin, and 5 ml of distilled water.

*Wet 1.0 g of carmine dye with 1.0 ml of concentrated NH_4OH, and dissolve this in 200 ml of distilled water. Add 10 g of commercial fibrin powder to the solution and let it stand overnight. Filter by suction. Wash thoroughly first with water, and finally with 0.3% acetic acid.

e. To tube #5 add carmine fibrin, 1 ml of 0.2% Na_2CO_3, and 5 ml of distilled water.

f. To tube #6 add carmine fibrin, 1 ml of pancreatin, and 5 ml of 0.4% HCl.

2. Carefully mix the tubes and incubate them in a 37°C waterbath. Observe the contents over the course of 2 hours.

STUDY QUESTIONS

1. Trypsin and chymotrypsin are two digestive enzymes of the pancreas that are active in the small intestine in the degradation of protein. Pepsin is a proteolytic enzyme of the stomach. What is the specific mode of action at the molecular level of each of these and how is this specificity achieved? What would be the effect of each enzyme on a peptide consisting of the following sequence of amino acids?

   NH_2-histidine—glutamic acid—tyrosine—threonine—lysine—histidine—glutamic acid—serine—arginine—aspartic acid—leucine—threonine—phenylalanine—COOH

   How might these specificities be used as analytical tools in the analysis of the primary structure of proteins?

2. Outline a set of analytical procedures you could use to identify the reaction product(s) from the activity of amylase on starch. List the main steps of your procedure as well as the detailed methodology you would use.

3. How are the peristaltic contractions of the small intestine initiated and regulated?

4. What are the specific causes of constipation and diarrhea? What medications and dietary measures can be employed to prevent or counteract each of these conditions?

REFERENCES

Anderson, S. 1973. Secretion of gastrointestinal hormones. *Ann. Rev. Physiol.,* 35:431.

Bortoff, A. 1972. Digestion: motility. *Ann. Rev. Physiol., 34:*261.

Brooks, F. P. 1970. *Control of Gastrointestinal Function: An Introduction to the Physiology of the Gastrointestinal Tract.* Macmillan, New York.

Christensen, J. 1971. The control of gastrointestinal movements: some old and new views. *New Eng. J. Med., 285:*85.

Davenport, H. W. 1972. Why the stomach does not digest itself. *Sci. Amer., 266(1):*86.

Davenport, H. W. 1971. *Physiology of the Digestive Tract,* 3rd ed. Year Book Medical Publishers, Chicago.

Gardner, J. D., M. S. Brown, and L. Laster. 1970. The columnar epithelial cell of the small intestine: digestion and transport. *New Eng. J. Med., 283:*1196.

Goldman, P. 1973. Therapeutic implications of the intestinal microflora. *New Eng. J. Med., 289*:623.

Gregory, R. A. 1968. Recent advances in the physiology of gastrin. *Proc. Roy. Soc., B170*:81.

McGuigan, J. E., and W. L. Trudeau. 1973. Differences in rates of gastric release in normal persons and patients with duodenal ulcer disease. *New Eng. J. Med., 288*:64.

McMinn, R. M. H. 1974. *The Human Gut.* Oxford Biology Reader, Oxford Univ. Press. (Available from Carolina Biological Supply Co.)

Meyer, J. 1975. *Human Nutrition,* 2nd ed. Charles C Thomas, Springfield, IL.

Rogers, T.A. 1958. The metabolism of ruminants. *Sci. Amer., 198*(2):34.

Stroud, R. M. 1974. A family of protein-cutting proteins. *Sci. Amer., 231*(1):74.

EXERCISE 15

Name _____

Laboratory Section _____ Date _____

RESULTS AND CONCLUSIONS

Salivary Digestion of Starch

Tube #	Lugol's Test	Benedict's Test
1		
2		
3		
4		
5		
6		

What conclusions can you draw about the action of salivary amylase?

Pancreatic Digestion of Fats

Tube #	Appearance	Odor
1		
2		
3		
4		

Account for the appearance of each of the tubes. Which tubes produce an odor? What substance produces the odor? Can you write a balanced chemical equation for the pancreatin reaction you have studied?

The Action of Bile Salts

Account for the action of bile salts. What is the detailed molecular basis for the action of bile? Can you write equations to illustrate this?

Exercise 15] Results and Conclusions

The Action of Rennin

Tube #	Appearance After 10 Minutes
1	
2	
3	
4	
5	

What is the effect of sodium oxalate on the activity of rennin? Does calcium chloride overcome the effect of oxalate? What can you conclude from this series of enzyme tests?

Digestion of Protein

| Tube # | Appearance After | | | | |
	60 minutes	75 minutes	90 minutes	105 minutes	120 minutes
1					
2					
3					
4					
5					
6					

Exercise 15] Results and Conclusions

Based on the experiments you have performed with gastric and pancreatic preparations, what can you conclude about the digestion of protein in the stomach and small intestine?

SECTION VI

Muscles

EXERCISE 16

Muscles: Some Elementary Considerations*

OBJECTIVES In this exercise we shall study several basic features of muscle anatomy (both gross and microscopic) and relate these to overall muscle functioning. Basic characteristics of striated, smooth, and cardiac muscle will be reviewed. We shall locate many of the motor points of our arm, hand, and finger muscles, and use these to study several elementary principles of muscle physiology.

There are at least 600 separate skeletal muscles in the human body. Many are essential for proper human movement, some are small and relatively unimportant. Most, if not all, are under voluntary control and can be moved at will.

Movement of the limbs and various other portions of the human body involves the action of many muscles working in highly concerted fashion. As a simple example, Figure 16-1 illustrates three muscles involved in flexion of the elbow joint in man. Contraction of the well-known biceps brachii plays a chief role in this movement; but the lesser known brachialis muscle is also considered a prime mover in elbow flexion, and the brachioradialis participates as well. Extension of the elbow joint is brought about by contraction of the triceps brachii, the antagonist of the three elbow flexors. In general, muscles that act in a manner opposed to one another (flexors/extensors; pronators/supinators, etc.) are termed antagonists. In every case, however, the only positive action of a muscle is to shorten its length as it contracts. The points of origin and insertion of a particular muscle determine the kind of movement that results from its contraction. Other examples are provided in Figure 16-2.

Each muscle of the body has specific points of origin and insertion. The origin is said to be the more or less immovable end and in the limb muscles is usually nearer to the trunk. The insertion is the farther end and moves as the muscle contracts. Muscles may attach to a bone in three ways.

(1) Direct attachment to the periosteum of the bone.
(2) Through a tendon, a tough cord of fibrous connective tissue.
(3) By an aponeurosis, a broad sheet of white connective tissue.

While most muscles have their origins and insertions in bone, some muscles, such as those in the skin and the eye, attach to cartilage or soft tissues.

*The first laboratory period devoted to the study of muscles affords a suitable occasion for viewing any of the several excellent films on muscles recommended at the end of this manual.

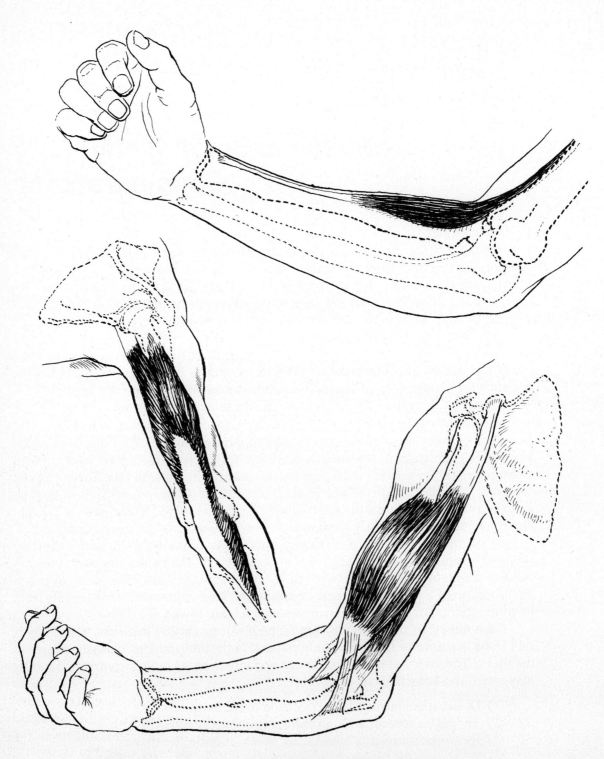

FIGURE 16-1 Flexor and extensor muscles of the elbow joint in man. *Top:* Brachioradialis. *Lower left:* Triceps and Anconeus. *Lower right:* Biceps brachii and Brachialis.

Exercise 16] Muscles: Some Elementary Considerations

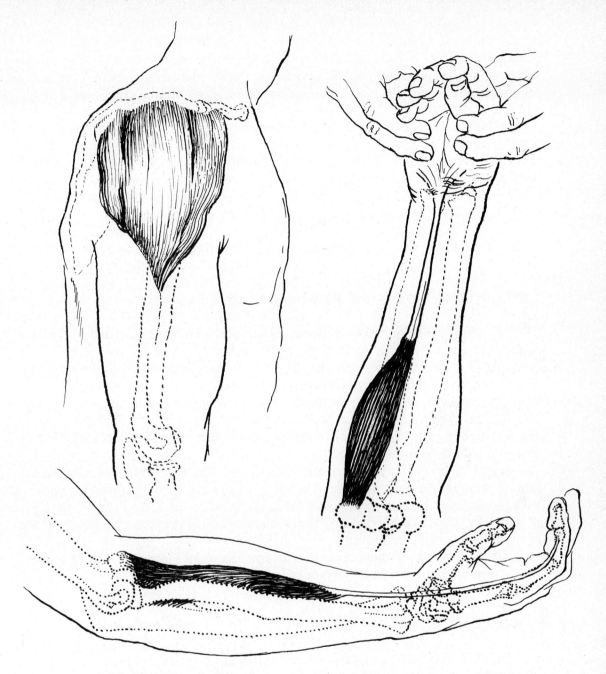

FIGURE 16-2 Three other muscles involved in human arm, hand, and finger movements. *Upper left:* Deltoid. *Upper right:* Palmaris longus. *Below:* Flexor digitorum.

In its fine structure, skeletal muscle consists of numerous muscle fibers usually running parallel to the long axis of the muscle. These fibers constitute the elementary units of muscle. A single muscle fiber may be 10-100 microns in diameter (approximately the range of diameters of a single human hair) and several millimeters to many centimeters in length. Individual muscle fibers, although they do not share protoplasmic connections with adjacent fibers, are bound together within common bundles (fasciculi). A muscle is formed from the systematic binding of these bundles by connective tissue.

The structure of individual muscle fibers is extremely complex, in that each fiber contains many nuclei not enclosed within specific cellular membranes. In addition, a given muscle fiber contains numerous subelements called myofibrils. The membrane surrounding a muscle fiber, the sarcolemma, is thicker and stronger than an ordinary cell membrane. If one equates the muscle fiber with a cell, this must be done with the firm reservation that this kind of "cell" is most unusual and atypical.

Individual muscle fibers, although physically separate, do act in coordinated fashion as so-called motor units. A motor unit consists of a single neuron and the muscle fibers it innervates. A ratio of 1 neuron to 50 muscle fibers is common, although the ratio may approach 1:10 in muscles playing a role in finely coordinated activity. In all cases, however, motor units obey the all-or-none law; that is, when a given motor unit is stimulated, all the muscle fibers in that particular unit contract maximally, or they respond not at all. Contraction of several or many of the motor units within a muscle produces movement of the bone onto which the muscle inserts.

THE LEVER ANALOGY FOR MUSCLE ACTION

An analogy may be drawn between muscle action and the mechanics of levers. Recall that there are three distinct classes of lever. A first class lever, of which the teeter-totter is a familiar example, has its fulcrum somewhere between the resistance (load) and the point of application of the force. In a second class lever (the wheelbarrow, or a nutcracker), the load is situated between the fulcrum and the force. A third class lever has its force point between the load and the fulcrum. A long-handled garden shovel illustrates this form of lever. Can you draw simple diagrams to illustrate each of these three types?

A moving body part serves, in effect, as a lever. The joint around which angular motion occurs is the fixed fulcrum. The force is provided by the contracting muscle, with the insertion serving as the point of application of the force. The resistance is provided by the pull of gravity, and the resistance point is the center of gravity of the portion of the body being moved.

Are all three types present in the human arm? Examine Figure 16-1 and determine which muscle(s) can serve as which type of lever.

BODY MOVEMENTS

The movements of the human body are extremely complex and varied. The nature of movement depends largely on the construction of the individual joint around which movement occurs and on the position of the muscles involved. Muscles working around a hinge type joint (such as the knee or elbow) are restricted in movement to a single plane, whereas muscles serving a ball-and-socket joint (such as the hip) can bring about more complex rotary movements. The different types of movement are as follows:

(1) *Flexion*. When the angle between two parts of a limb is decreased, the limb is said to be flexed. Flexion takes place at the elbow joint when the forearm is moved toward the upper arm. The term *flexion* may also be applied to movement of the head against the chest and the thigh against the abdomen.

(2) *Extension*. Increasing the angle between two portions of a limb or parts of the

body is termed extension. This type of movement is exactly the opposite of flexion. Straightening the arm from a flexed condition is an example of extension.

(3) *Abduction.* The movement of a limb away from the median line of the body is abduction. This term may be applied to the fingers and toes, using the longitudinal axis of the limbs as a point of reference.

(4) *Adduction.* When a limb is moved from an outward position toward the median line, the movement is termed adduction. Adduction and abduction are opposite types of movement. As in the case of abduction, adduction can also be applied to the fingers and toes by using the longitudinal axes of the limbs as points of reference.

(5) *Circumduction.* Rotational movement of a limb in a ball-and-socket joint in such a manner that the distal portion of the limb describes a circle is circumduction.

(6) *Supination.* Movement of the palm of the hand and the forearm upward by rotation of the radius about the ulna is supination. As a result of this movement, the radius and the ulna become more or less parallel to each other.

(7) *Pronation.* The palm of the hand is moved from an upward facing position to a downward facing position during pronation. In this case, the radius crosses over the ulna. Pronation and supination are opposite types of movement.

Procedures

Study a few illustrative examples of these movements as follows: Work in groups of three. One student serves as the subject. The second will apply resistance to the areas of the body indicated below. The third student will palpate the muscle(s) being tested. Detailed diagrams of the human musculature may be needed to identify several of the muscles.

1. *Deltoid.* The subject attempts to abduct the laterally raised arm against resistance.
2. *Trapezius,* upper portion. The subject attempts to elevate the shoulder against resistance.
3. *Latissimus dorsi.* The subject attempts to adduct the arm from an extended and lateral position against resistance.
4. *Biceps.* The subject attempts to flex the supinated forearm against resistance. Compare this with the situation in which the subject flexes his own biceps without external resistance. Palpate the triceps in both cases.
5. *Triceps.* The subject attempts to extend the forearm which is flexed at the elbow.
6. *Extensor pollicis longus.* The subject extends the thumb against resistance.
7. *Quadriceps.* An attempt is made to extend the knee against resistance on the leg.
8. *"Hamstring" muscles.* The subject is prone and attempts to flex the knee against resistance.

MICROSCOPIC ANATOMY OF HUMAN MUSCLE TISSUE

Examine prepared slides of human muscle tissue to familiarize yourself with the microscopic anatomy of skeletal, smooth, and cardiac muscle.

Materials

prepared slides of human skeletal muscle (cross-sectional and longitudinal sections)
prepared slides of human smooth and cardiac muscle
prepared slides of motor end plate and muscle spindle
microscopes
stage and ocular micrometers
drawing paper
pencils

Procedures

1. Carefully study under high power a prepared slide of human skeletal muscle in longitudinal section. Can you identify individual muscle fibers? Compare these same structures in a cross section. Make a careful three-dimensional drawing of a portion of skeletal muscle. Identify nuclei, the sarcolemma, myofibrils. Can you see fasciculi? What is the width of a single muscle fiber? If suitable micrometers are not available, place a single hair on the slide and compare the dimensions of a muscle fiber with those of a hair. Under oil immersion carefully study the pattern of striations within a muscle fiber. Draw accurately and label the various bands.
2. Make detailed observations on prepared slides of smooth and cardiac muscle. How do these compare and contrast with skeletal muscle?
3. Make a careful drawing of a motor end plate. Draw and label a muscle spindle. Handle these slides very carefully as they are very expensive.

LOCATION OF MOTOR POINTS IN MAN

Isolated nerve-muscle preparations (such as the well-known gastrocnemius muscle–sciatic nerve from the frog) are often used in basic physiological study. The isolated muscle can be stimulated with electrodes attached directly to the muscle itself or indirectly through its nerve. Many principles of muscle physiology can also be studied in human subjects using very mild electrical stimulation.* In this part of the exercise we shall locate the motor points for many of our own arm, hand, and finger muscles, and use this knowledge to study several important principles of muscle physiology.

The motor point is taken to represent the point at which the motor nerve enters the muscle, and it is usually located over the central "belly" of the muscle. Electrical stimulation at the motor point causes contraction of the muscle. Clinical motor point testing is a routine diagnostic procedure for suspected neuromuscular damage. Electrical stimulation is also used therapeutically to stimulate denervated muscles in patients with poliomyelitis and other conditions where nerve damage and consequent

---

*Lest students suffer anguish over this kind of experiment, it is to be emphasized that under ordinary circumstances motor point stimulation produces quite mild sensations. Any student who does not wish to take part in this portion of the exercise is free to refrain, as pointed out in the Preface. Also, the class instructor should take careful note of any precautions specified by the manufacturer of the electrical stimulators. The voltage setting should be at zero while the instrument is being attached to the subject, and no other electrical apparatus should be attached to the student while the stimulators are being used.

muscular inactivity produce a loss of tone and eventual atrophy of the muscle. Other therapeutic techniques are passive movement and massage.

Materials

EKG gel (or 4% saline solution)
electrical stimulator (see Note)
reference electrode (such as the plate type electrode often used in EKG determination)
elastic straps for securing plate electrodes
exploring electrode (the electrode supplied with the stimulator) lead wires for electrodes

Note: Any square wave generator, such as those used in frog muscle experiments, is acceptable (e.g., Harvard Instrument 344 or Phipps and Bird 611). The instructor must verify that the stimulator is designed for use on humans and exercise all precautions specified by the manufacturer.

Procedures

1. Connect the two electrodes to the stimulator. The exploring electrode is attached to the red terminal (STIM), the reference electrode to the black terminal (GND). (If the stimulating electrode is a two-lead, two-probe type, determine which of the two probes is active and connect this one to the STIM terminal. Leave the second of the two leads disconnected.) Soak the stimulating electrode in saline solution for a few minutes.

2. Moisten the underside of your arm with EKG solution (or 4% saline) and secure the plate electrode in position with the elastic arm band. Rest your arm comfortably on the laboratory table. Rub liberal amounts of EKG gel over the portion of your arm to be explored.

3. Before you turn on the stimulator, be certain that the voltage adjustment and the duration switch are turned down. Familiarize yourself with the dials on the face of the stimulator. Now turn on the stimulator and set the stimulating frequency control at one impulse per second. Turn the voltage to about 60 volts. (Rest assured that the current flow from the instrument is extremely small.) Use the diagram of the human arm depicted in Figure 16-3 showing the approximate position of the various motor points as a basis for exploration of your own arm. How many of the motor points can you locate? Can you identify flexors and extensors? Can you locate pronators and supinators? Figure 16-3 illustrates several (but not all) of the motor points in the human arm.

STUDY QUESTIONS

1. Prepare a comprehensive chart showing the anatomical and physiological similarities and differences among striated, smooth, and cardiac muscle.

2. What is a muscle spindle, and how does it function in muscle action? Draw a diagram to illustrate your answer and discuss the pertinent experimental evidence.

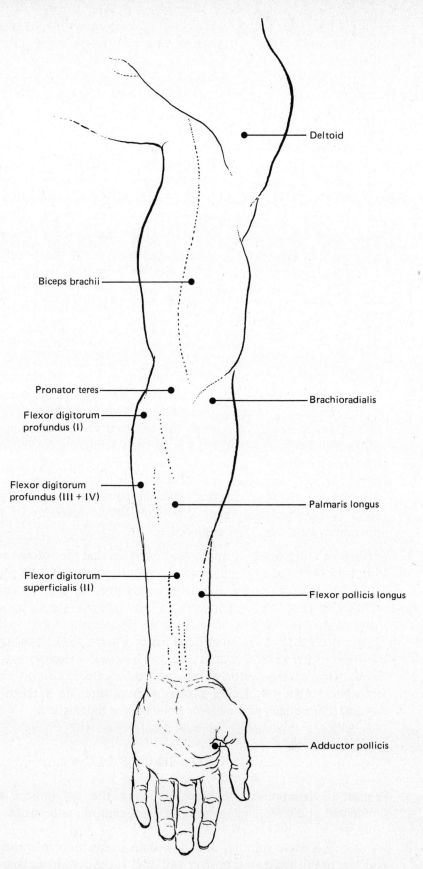

FIGURE 16-3 Motor points for several muscles in the human arm.

3. The sliding filament model is a widely accepted theory of muscle contraction. Outline this theory, and cite the experimental evidence in support of it.

4. The sliding filament theory makes a very specific suggestion about the role of ATP in muscle contraction. What is it, and what is the postulated role of creatine phosphate in muscle contraction?

5. Many textbooks of biology show a diagram of the so-called motor area of the cerebral cortex in man. This area has been defined by investigators such as Dr. Wilder Penfield of the Montreal Neurological Institute through electrical stimulation of the brain during operations on epileptic patients while the patient is fully conscious. The surgeon stimulates various regions of the brain while an observer questions the patient and notes the responses. In this way a map of the motor areas of the entire human body is systematically constructed. What is the actual experimental evidence that the motor area functions in this manner during an ordinary act of will; i.e., that the act of cerebration initiates the transmission of nerve impulses along a motor pathway and that this results in the contraction of particular muscles of the body?

REFERENCES

Bourne, G. H. (ed.). 1972. *The Structure and Function of Muscle,* 2nd ed. (4 vols.). Academic Press, New York.

Close, R. I. 1972. Dynamic properties of mammalian skeletal muscles. *Physiol. Rev., 52*:129.

Evans, F. G. 1961. *Biochemistry Studies of the Musculoskeletal System.* Charles C Thomas, Springfield, IL.

Evarts, E. V. 1973. Brain mechanisms in movement. *Sci. Amer., 229*(1):96.

Gergely, J. (ed.). 1964. *Biochemistry of Muscle Contraction.* Little, Brown, Boston.

Hubbard, J. I. 1973. Microphysiology of vertebrate neuromuscular transmission. *Physiol. Rev., 53*:874.

Huxley, H. E. 1965. The mechanism of muscular contraction. *Sci. Amer., 213*(6):18.

Huxley, H. E. 1969. The mechanism of muscular contraction. *Science, 164*:1356.

Kendall, H. O., et al. 1971. *Muscles: Testing and Function,* 2nd ed. Williams and Wilkins, Baltimore.

Merton, P. A. 1972. How we control our muscles. *Sci. Amer., 226*(5):30.

Murray, J. M., and A. Weber. 1974. The cooperative action of muscle proteins. *Sci. Amer., 230*(2):58.

Rasch, J., and R. Burke. 1963. *The Science of Human Movement,* 2nd ed. Lea and Febiger, Philadelphia.

Snell, F. 1965. *Biophysical Principles of Structure and Function.* Addison-Wesley, Reading, MA.

Weber, A., and J. M. Murray. 1973. Molecular control mechanisms in muscle contraction. *Physiol. Rev., 53*:612.

EXERCISE 16

Name _____

Laboratory Section _____   Date _____

RESULTS AND CONCLUSIONS

Human Muscles and the Lever Analogy

Are all three classes of levers present in the human arm? Which classes are?

Body Movements

For each of the muscles you study by palpation, prepare an accurate drawing showing the points of origin and insertion. What movements does each muscle control?

Microscopic Anatomy

Prepare labeled drawings of the microscopic anatomy of striated, smooth, and cardiac muscle; indicate the dimensions wherever possible. Make a careful drawing of skeletal muscle showing a motor end plate, and label the appropriate features. Also draw a muscle spindle.

Motor Point Determination

Summarize and discuss your findings on the motor points on your own arm. Prepare an outline drawing to show the motor points you were able to locate. What is the threshold for stimulation? Does repeated stimulation enhance contraction?

EXERCISE 17

Principles of Muscle Physiology

OBJECTIVES In this exercise we shall study basic muscle physiology using in vivo stimulation of our own arm and finger muscles. The effects of varying stimulus strength and frequency will be investigated, as well as the phenomena of tetanus and fatigue. The finger ergometer and hand dynamometer will be used to study other basic aspects of muscle physiology.

Many principles of basic muscle physiology may be studied in vivo with the apparatus illustrated in Figure 17-1. In this setup the square wave generator is used to stimulate the motor point of one or more of the finger flexors.* As the muscle contracts, the movements are recorded. With this apparatus, you can produce a simple muscle twitch and investigate the threshold for stimulation, summation, the Treppe ("staircase") phenomenon, tetanus, fatigue, and other related phenomena.

MEASUREMENT OF FINGER FLEXION

Materials

EKG gel (or 4% saline solution)
electrical stimulator (1 per two to four students)**
reference electrode (such as the plate-type electrode often used in EKG determination)
exploring electrode (the electrode supplied with the stimulator)
elastic straps for securing plate electrodes
lead wires for electrodes
suitable recording equipment (see Note)
ringstands
small pulleys
inelastic cord or wire

---

*Flexion of the fingers of the human hand, exclusive of the thumb, is effected by many muscles, including several termed flexor digitorum. A series of four flexor digitorum subliminis muscles insert into the bases of each of the distal finger bones (phalanges). Another series of four flexor digitorum superficialis muscles attach to the middle phalanges. Roman numerals I-IV designate each of the four fingers starting with the index finger. See Figure 16-2.
**The instructor should be certain to read the footnote on page 194 and the Note on page 195 before conducting this exercise.

Note: The Harvard chart mover (model 450) equipped with the Isotonic Muscle Contraction Module (model 270) and Event/Timer Marker Module (model 283) or standard kymographs are acceptable.

Procedures

1. Set up the apparatus as shown in Figure 17-1. The plate electrode should be attached to the undersurface of your arm or to your ankle over a fairly liberal amount of the EKG gel and secured firmly with an elastic strap. Rest your arm comfortably on the laboratory bench in a supine position. Follow the same instructions and precautions in the use of the square wave stimulator as previously described in Exercise 16 for motor point determination. Locate the motor point for flexion of the middle and/or ring finger. A stimulating frequency of 1 impulse/second and a voltage setting of 60–75 volts should produce movement. After you have located the motor point, tie the inelastic cord to your finger, pass it over the pulley system, and attach it to the recording device. (If you use the Harvard chart recorder equipped with the isotonic contraction module, it is necessary to add a small counterweight to the arm of the module; this is not shown in Figure 17-1.) Induce flexion with the stimulator and check the apparatus to see that you are making a suitable record of the muscle contraction.
2. Set the stimulator below the threshold for a detectable response (less than 50 or 60 volts) at a frequency of 1 impulse/second. Gradually increase the voltage until contraction occurs.
3. Select a voltage setting slightly below the threshold and slowly increase the impulse frequency. What results are obtained? Repeat this experiment at a voltage setting fairly well above the threshold value. Obtain clear recordings of muscle contraction at 1, 2, 3, 4, 5, 10, and 25 impulses/second.
4. Can you design and execute a simple experiment to measure the *latent period* for muscle contraction? What limitations are present in the experimental setup itself? How could these be overcome?
5. Can you design and execute a series of simple experiments to measure *chronaxie* and *rheobase* using the experimental setup you have established? You will need a square wave generator (such as the Phipps and Bird 611 stimulator depicted in Figure 17-1) with which the *duration* of the stimulus can be varied.

MUSCLE FATIGUE AND TETANUS

When a muscle is caused to contract repeatedly with very short intervals of rest, a decrease in the response can be seen in both excised and intact muscle. After prolonged contraction with both isometric and isotonic exercise, the ability of the muscle to contract disappears altogether and the muscle is said to be fatigued. An isolated frog gastrocnemius muscle stimulated through the sciatic nerve will no longer contract after repeated stimulation, but the muscle itself still responds if the stimulating electrode is applied directly to it. This finding, together with other results showing that the motor nerve is still able to function after prolonged stimulation, have led to the conclusion that the *site of fatigue* in the isolated frog gastrocnemius muscle is in the myoneural junction. The site of fatigue in intact human muscle is a highly complex and as yet unresolved issue. Several references at the end of this Exercise summarize recent experimental findings.

Exercise 17] Principles of Muscle Physiology

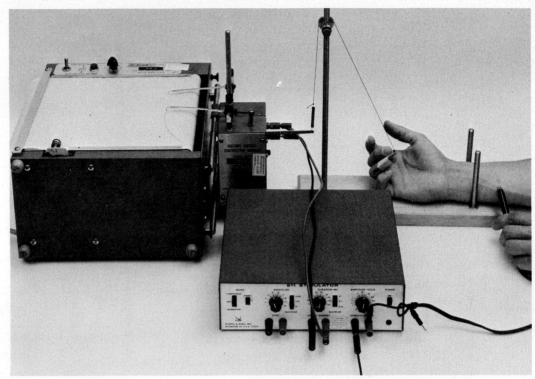

FIGURE 17-1 Apparatus for stimulating the flexor digitorum muscle and recording the response.

Materials

finger ergometer (see Note)
suitable recording apparatus (e.g., Harvard chart recorder)
hand dynamometer
sphygmomanometer
stethoscope
metronome
containers for hot and cold water

Note: A reliable finger ergometer is supplied by Phipps and Bird, Inc., P. O. Box 27324, 8741 Landmark Rd., Richmond, VA 23261.

Procedures

Finger Ergometry
1. Set up the finger ergometer and the recording apparatus as illustrated in Figure 17-2. It is necessary to counterbalance the lever arm of the isotonic contraction module with a small weight. It is also helpful to secure the ergometer to the benchtop with suitable clamps. Adjust the tension of the spring for a particular subject so that the trigger can be pulled to the maximum extent possible. The stroke adjustment may also be regulated.
2. The subject should seat himself comfortably with his arm resting on the tabletop and his wrist positioned between the verticle support bars. The metronome is set

204 Section VI] Muscles

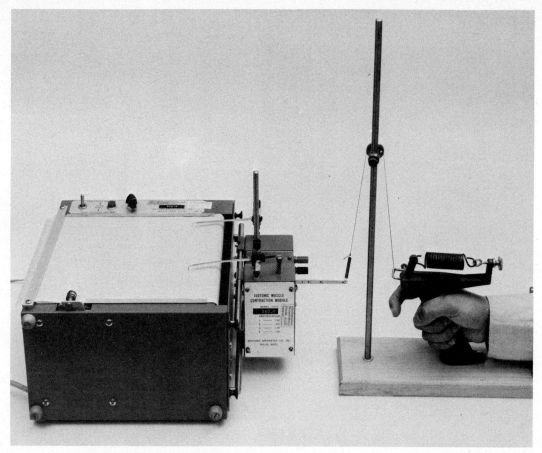

FIGURE 17-2 The finger ergometer and recorder for in vivo measurement of isotonic muscle contraction and fatigue.

at 60 beats/minute and at time 0 the subject starts to pull the trigger at a rate of 1 pull per second. The subject continues to pull according to the cadence set by the metronome—the lab partner should watch this closely—until the amplitude of the pull is half its original height. This may take 5-10 minutes. The subject then rests for 30 seconds, and repeats the experiment exactly.

3. Repeat the basic experiment outlined above under the following conditions.
   a. With circulation to the arm occluded with a blood pressure cuff inflated to diastolic pressure (to be determined with the sphygmomanometer and stethoscope).
   b. With circulation to the arm occluded to 200 mm Hg (i.e., above systolic pressure).
   c. After the subject's arm has been immersed in 60°C water for 10 minutes.
   d. After the subject's arm has been immersed in 10°C water for 10 minutes.
   Record the results.

Hand Dynamometry

The purpose of this experiment is to compare the rate of fatigue in isometric contraction with that of rhythmic (isotonic) work using the hand dynamometer (Figure 17-3). These parameters will also be measured under conditions of circulatory occlusion.

1. Each student measures his grip strength in each of his two hands and uses the maximum values to compute his fatigue point (i.e., half maximum grip strength).
2. Students should work in groups of three to measure rates of fatigue. For isometric exercise, the subject squeezes the dynamometer continuously. The second student calls out the strength of contraction every 2 seconds until the fatigue point is reached. The third student acts as data recorder. The fatigue rate is determined for the opposite hand and the determinations can also be repeated with the circulation occluded to 200 mm Hg. Graph the results. What conclusions can you draw?
3. For rhythmic (isotonic) exercise, the subject squeezes the dynamometer maximally, alternating one-second contraction, one-second relaxation until fatigue (half maximal strength) is reached. The second student reads the strength of each individual contraction, and the third student records. These measurements can be repeated with circulations occluded to 200 mm Hg. Graph the results. How do the results for isometric and isotonic exercise compare?

STUDY QUESTIONS

1. Karpovich and Sinning, in their textbook *Physiology of Muscular Activity*, mention six possible sites of muscle fatigue in man as follows: (1) the muscle fiber itself, (2) the nerve motor end plate, (3) the motor nerve fiber, (4) synapses within

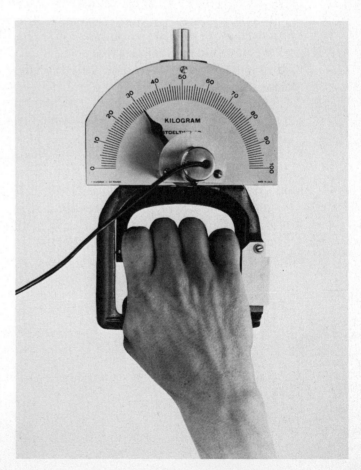

FIGURE 17-3   The hand dynamometer.

the nerve ganglia and the central nervous system, (5) the nerve cell body, and (6) the end organs of sense in the muscle and elsewhere in the body. Which of these do the authors favor as the probable seat of muscle fatigue in man, and why?

2. What factors determine the strength of a muscle? What are the advantages of isotonic versus isometric training methods? What is interval training, and what is its physiological basis?

3. What are the biochemical and histological changes that occur in muscle tissue with aging? With disuse and consequent atrophy?

4. What is the experimental evidence that recruitment of additional motor units in muscle is caused by increased motor nerve activity in vivo?

REFERENCES

Bartley, S. 1962. *Fatigue: Mechanism and Management,* 2nd ed. McGraw-Hill, New York.

deVries, H. 1966. *Physiology of Exercise for Physical Education and Athletics.* Wm. C. Brown Co., Dubuque, IA.

deVries, H. 1971. *Laboratory Experiments in Physiology of Exercise.* Wm. C. Brown Co., Dubuque, IA.

Doss, W. S., and P. V. Karpovich. 1965. A comparison of concentric, eccentric and isometric strength of elbow flexors. *J. Appl Physiol., 20*:351.

Karpovich, P. V., and W. E. Sinning. 1971. *Physiology of Muscular Activity,* 7th ed. W. B. Saunders, Philadelphia.

Katz, B. 1966. *Nerve, Muscle and Synapse.* McGraw-Hill, New York.

Merton, P. A. 1972. How we control the contraction of our muscles. *Sci. Amer., 226*(5):30.

Moudgil, R., and P. V. Karpovich. 1969. Duration of a maximal isometric contraction. *Res. Quart., 40*:3.

Pernow, B., and B. Saltin. 1971. *Muscle Metabolism During Exercise.* Plenum, New York.

Rasch, P. J., and L. E. Morehouse. 1957. Effects of static and dynamic exercise on muscular strength and hypertrophy. *J. Appl. Physiol., 11*:29.

Royce, J. 1958. Isometric fatigue curves in human skeletal muscle with normal and occluded circulation. *Res. Quart., 29*:204.

Simonsen, E 1971. *Physiology of Work Capacity and Fatigue.* Charles C Thomas, Springfield, IL.

EXERCISE 17

Name _____

Laboratory Section _____ Date _____

RESULTS AND CONCLUSIONS

Electrical Stimulation of Finger Flexors

Attach your kymograph or recorder charts to your report to illustrate your findings. Summarize your results by answering the following questions. What is the threshold value in volts for muscle contraction? At what stimulating frequency does sustained muscle contraction occur? What is the *Treppe* ("staircase") phenomenon? What is temporal summation? What is spatial summation? Did you demonstrate summation in your experiments? If so, which type(s)? What is tetanus? Is maximal contraction ever obtained, even at high frequency and fairly high voltages? Why? Do the small muscle responses seen at low voltages violate the all-or-none law? Explain.

Describe and interpret your results on the latent period for muscle contraction, chronaxie, and rheobase.

Finger Ergometry

Attach the charts to your report to illustrate your findings with the finger ergometer. How do the results compare for the two consecutive series of pulls? What muscles are involved in this activity? What are the subjective feelings of fatigue?

Discuss the effects of circulation occlusion and temperature on fatigue.

What simple observation demonstrates tetanus? How could you show this with a rubber ball?

Hand Dynamometry

Compile the class data for this part of the exercise and compare them in the following ways.

(1) How does the absolute grip strength in the dominant hand compare with the nondominant hand?
(2) How does relative endurance (fatigue rate) compare in the dominant and nondominant hands?

(3) How can isotonic and isometric fatigue curves be compared?
(4) What are the effects of circulatory occlusion on the two types of exercise?

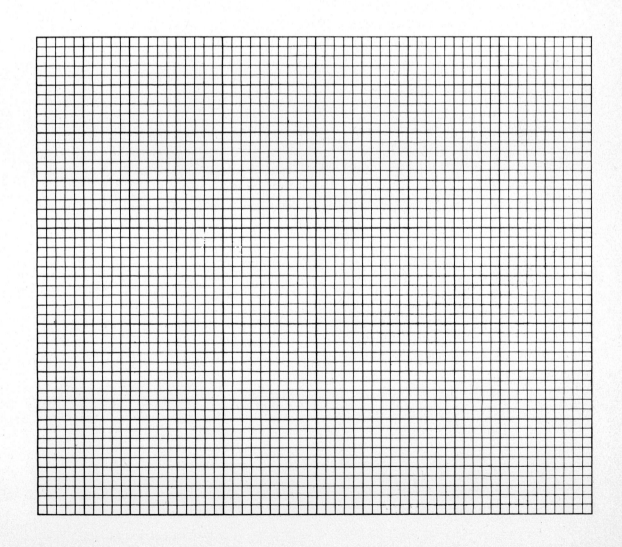

SECTION VII

Cellular and Subcellular Processes

EXERCISE 18

Osmotic Phenomena and Cell Permeability

OBJECTIVES In this exercise we shall study elementary aspects of the theory and methodology for determining the osmotic properties of a solution. Living cells, including the leaf cells of *Elodea* plants and human erythrocytes, will be studied as osmometers. Spectrophotometric tests for measuring the rate of penetration of various solutes into erythrocytes are to be carried out.

The properties that enable the cell membrane to regulate the exchange of materials into and out of the cell are of prime importance for the maintenance of cellular life. The movement of water and dissolved substances through the cell membrane provides a supply of oxygen and nourishment, the elimination of waste materials, and a proper intracellular environment, all of which are essential to the well-being of cells and tissues.

In many instances the permeation of solute molecules into the cell is solely a function of concentration gradients; a substance present in high concentration outside the cell moves in until the intra- and extracellular concentrations are equal. Exit flow is simply a reversal of the process, i.e., a high intracellular concentration moves outward. Detailed studies have shown that among the factors influencing the flow of solutes through the membrane are the molecular weight, fat solubility, and ionic character of the substance. Smaller molecular weight compounds with high lipid solubility readily pass through the cell membrane. Membranes are generally quite permeable to anions of inorganic electrolytes, but relatively impermeable to cations of the same electrolytes.

In addition to processes of passive entry, there are also active permeation mechanisms that work *against* a concentration gradient; that is, many substances are brought into a cell even though the intracellular concentration is higher. Such a process requires the expenditure of energy and is called active transport. Specific carrier mechanisms (or "permeases") for these substances are located in the cell membrane.

Water passes into and out of cells through the cell membrane, but this movement cannot be detected if solute concentrations on both sides of the membrane are the same. Only when the solute concentrations are unequal can a net movement of water be observed. This flow continues until the internal and external concentrations have been equalized. In many cases, the membrane is impermeable to various solutes, and water is the main substance moving freely through the membrane. In other circumstances, solute molecules move through the membrane at rates depending on the chemical nature of the substances, and the specific properties of the membrane.

The diffusion of water through a semipermeable membrane is termed osmosis. Osmosis can be demonstrated with a simple device called an osmometer, such as the one depicted in Figure 18-1. A solution of a given substance is placed within a thistle tube and separated from distilled water by a semipermeable membrane. In the simplest case, a solute is chosen that does not penetrate the membrane at all. As water diffuses through the membrane into the thistle tube, the internal volume increases and the fluid level in the tube rises. The pressure exerted by the solution is proportional to the height of the fluid in the column (h) and is termed the osmotic pressure (π). Water continues to enter the thistle tube until the downward pressure exerted by the column of the fluid is exactly equal to the pressure of water entering the tube through the membrane. In this case $\pi = h \times k$, where h is the height of the column and k is a proportionality constant depending on temperature and the concentration of the solute.

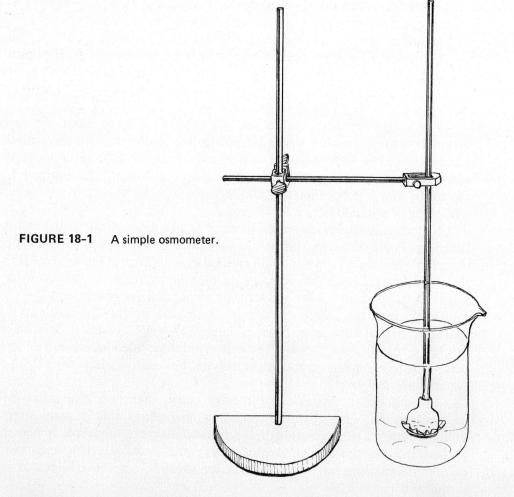

FIGURE 18-1 A simple osmometer.

Simple measurements have shown that the osmotic pressure exerted by a solution is directly proportional to its concentration. In simple terms, if you double the concentration of a solution, you double its osmotic strength. The physical chemist J. T. van't Hoff has demonstrated that this finding can be related in an important way to the well-known behavior of gases. The ideal gas law states that

$$PV = nRT \tag{1}$$

P = pressure in atmospheres
V = volume of gas in liters
n = number of moles of gas
R = universal gas constant (0.082 liter-atm/mole-°K)
T = absolute (Kelvin) temperature

At 273°K and 1 atm pressure, one mole of gas occupies a volume of 22.4 liters. If this volume of gas is compressed to 1 liter, it now exerts a pressure of 22.4 atmospheres. The correspondence between the behavior of gases and that of solutions is based on the comparison of one mole of gas compressed to a liter with one mole of a solid substance dissolved in a liter of solvent. The compressed gas exerts an atmospheric pressure of 22.4 atm; the substance in solution exerts an osmotic pressure of 22.4 atm. Empirical studies have validated this relationship within certain limits.

Exact derivation of the equation for osmotic pressure is as follows. Equation 1 can be rewritten as

$$P = \frac{r}{V} \times RT \tag{2}$$

Now, r/V for a solution is the number of moles of solute per volume of solvent (in this case 1 liter), and this is simply the molal concentration. Equation 2 now becomes

$$P = CRT \quad \text{or} \quad \pi = CRT \tag{3}$$

where C is the molal concentration of the solution. For an ideal situation, the osmotic pressure π of a 1 molal solution is

$$\pi = 1\frac{\text{mole}}{\text{liter}} \times 0.082 \frac{\text{liter-atm}}{\text{mole-°K}} \times 273°K = 22.4 \text{ atm}$$

Living cells contain semipermeable membranes and act as tiny osmometers. Sea urchin eggs, aquatic plant cells, and mammalian red blood cells take up water when placed in *hypotonic* solutions—that is, solutions having a lower osmotic strength than the internal cellular contents. These same cells lose water when placed in more concentrated *hypertonic* solutions. A medium in which neither shrinkage nor swelling takes place is said to be *isotonic* to the cell contents.

USE OF THE OSMOMETER IN STUDYING OSMOTIC PHENOMENA

Materials

simple osmometers (5 per group of four students; see Note)
250 ml beakers or cups supplied with the osmometers (5 per group)
10 ml pipets (3 per group)
10% sucrose solution (250 ml per group)
20% sucrose solution (250 ml per group)

Erlenmeyer flasks (2 per group)
distilled water
small rulers graduated in mm
wax marking pencils

Note: Inexpensive osmometers are available from Carolina Biological Supply Co.

Procedures

1. The membranes in the osmometers should be thoroughly wetted for at least 30 minutes before the start of the experiment. Membrane leaks can be detected by pipetting a dilute solution of the high molecular weight dye, Congo red, into the thistle tube.
2. Solutions are pipetted into the osmometer as follows:
   a. Osmometer #1 receives distilled water in the thistle tube, distilled water in the beaker. Be certain that there are no trapped air bubbles in the thistle tube. Gentle tapping should release them. The initial height of the water in the tube should be set at an appropriate height so that water movement can be detected in either direction. A height of 10–20 cm above the beaker is suggested. Mark the tube at the meniscus of the fluid with the wax pencil.
   b. Osmometer #2 receives 10% sucrose solution in the thistle tube, distilled water in the beaker. Set the fluid height as suggested above and mark the tube.
   c. Osmometer #3 receives 20% sucrose solution in the thistle tube, distilled water in the beaker. Set the water height as suggested above and mark the tube.
   d. Osmometer #4 receives distilled water in the thistle tube, 10% sucrose solution in the beaker. Set the fluid height about 20 cm above the beaker and mark the tube.
   e. Osmometer #5 receives distilled water in the thistle tube, 20% sucrose in the beaker. Set the fluid height about 20 cm above the beaker and mark the tube.
3. Carefully measure the height of the fluid in each of the osmometers at 10 minute intervals for 1 or 2 hours, and record the data. Plot graphs of the rate and direction of flow in each case.

LIVING CELLS AS OSMOMETERS

Materials

human blood (3 ml per two students; see Note)
fresh *Elodea* plants (1 bunch per laboratory section)
clinical centrifuges and centrifuge tubes
test tube racks (1 per two students)
1 ml pipets (2 per two students)
5 ml pipets (2 per two students)
small forceps
2% stock NaCl solution (25 ml per two students)

Exercise 18] Osmotic Phenomena and Cell Permeability

2 *M* stock sucrose solution (10 ml per two students)
distilled water
125 or 250 ml Erlenmeyer flasks (1 per two students)
microscopes
microscope slides and coverslips
wax marking pencils

Note: Outdated human blood or blood unacceptable for transfusion can be purchased very inexpensively from local blood banks.

Procedures

Erythrocytes
1. Set up a series of tubes as follows:

Tube #	2% NaCl	Distilled Water	Final Salt Concentration
1	0	5 ml	0
2	0.25 ml	4.75 ml	0.1%
3	0.50 ml	4.5 ml	0.2%
4	0.75 ml	4.25 ml	0.3%
5	1.0 ml	4.0 ml	0.4%
6	1.25 ml	3.75 ml	0.5%
7	1.5 ml	3.5 ml	0.6%
8	1.75 ml	3.25 ml	0.7%
9	2.0 ml	3.0 ml	0.8%
10	2.25 ml	2.75 ml	0.9%
11	2.51 ml	2.5 ml	1.0%
12	5.0 ml	0	2.0%

2. To each of the tubes add 1 drop of blood and mix the contents thoroughly. Allow the tubes to incubate on the benchtop and examine them periodically with the microscope for signs of hemolysis. Hemolysis may occur in some of the tubes immediately and in others much later. For this reason it is advisable to set up this experiment early in the laboratory period.
3. After an incubation period of 2 hours, prepare microscope slides from several tubes and examine the supernatants for signs of hemolysis. At what point is hemolysis first evident? At what concentration(s) is hemolysis complete?

Elodea
1. Remove a small *Elodea* leaf from the growing tip of the plant, mount it in a drop of aquarium water on a slide, carefully add a coverglass, and observe it under the high power of the microscope. Be certain to close the microscope diaphragm to decrease the illumination. Carefully draw what you see. How many cell layers thick is the leaf? Can you identify the cell wall and the cell membrane? What organelles do you see? Do you observe cyclosis?
2. Place a freshly blotted leaf on a slide and cover it with a coverglass. While viewing the leaf under high power of the microscope, ask your partner to pipet a small volume of 2 *M* sucrose onto the slide at the edge of the coverglass. Carefully observe a few *Elodea* cells. Can you observe plasmolysis? Draw what

you see. After you have finished your drawing, flush distilled water under the coverglass and watch carefully for any signs of deplasmolysis. Describe your observations.

ERYTHROCYTE PERMEABILITY

The penetration of substances into red blood cells can be studied with great precision. A suspension of erythrocytes is ordinarily very turbid and scatters a beam of light shone through it. If these cells are placed in hypotonic medium, they take up water and burst (hemolysis). The resulting mixture of red cell "ghosts" and soluble intracellular proteins (largely hemoglobin) is quite clear and transmits a high proportion of incident light. Accordingly, a quantitative study can be conducted by following the time course of hemolysis in a spectrophotometer at wavelength 600 nm. Hemoglobin absorbs very little light at this wavelength, but the decrease in light scattering due to hemolysis can be carefully monitored.

The penetration of organic solutes into red cells can be measured in the same manner. If a given solute enters the cell, this causes the osmotic strength of the intracellular contents to increase. Water then enters and hemolysis occurs.

Materials

Bausch & Lomb Spectronic 20 spectrophotometers (or a suitable alternative) (1 per four students)
spectrophotometer tubes (4 per two students)
human blood (5 ml per two students)
parafilm squares
test tubes (12 per two students)
test tube racks (1 per two students)
5 ml pipets (10 per two students)
Pasteur pipets (10 per two students)
stopwatches (or wristwatches with a second sweep)
glycerol solutions: 0.3, 0.5, 0.7, and 1.0 M (10 ml per two students)
0.3 M urea solution (10 ml per two students)
0.3 M glucose solution (10 ml per two students)
0.3 M monoacetin (10 ml per two students)
0.3 M diacetin solution (10 ml per two students)
0.3 M triacetin solution (10 ml per two students)

Procedures

Penetration of Glycerol
1. The basic procedure for studying penetration is as follows: Set the spectrophotometer to wavelength 600 nm. Pipet 5.0 ml of the solution to be tested into a clean spectrophotometer tube, place the tube into the compartment of the photometer, and adjust the transmittance to 100% (absorbance at 0). This procedure calibrates the machine. Now remove the tube and place it in a test tube rack.

2. Your partner will add exactly 1 drop of blood to the tube. The timing watch starts when the drop hits the solution. Quickly mix the contents of the tube by inverting it with a parafilm strip. *Speed is of the essence!* Place the tube in the photometer and begin immediately to make absorbance readings at 5 second intervals. With a little practice you should be able to make your first reading at $t = 10$ seconds. The student reading the spectrophotometer should call out the absorbance value at each 5 second interval and a second student should record the data. You may find it convenient to have a third class member call out the time intervals.
3. After you have practiced this procedure, measure the time course of hemolysis with the four glycerol solutions (0.3, 0.5, 0.7, and 1.0 M). Graph the results.

Penetration of Other Substances
1. Using the basic procedures outlined above, study the rate of penetration of urea, glucose, and the various acetylated glycerols (mono-, di-, and triacetin). Several of these substances penetrate very rapidly, and it will be necessary to make readings a few seconds after adding the drop of blood. One possible modification of the protocol is to add the drop of blood directly to the tube inside the spectrophotometer and then mix the contents very rapidly with a glass rod.

STUDY QUESTIONS

1. Are the terms isotonic and isosmotic synonomous? (*Hint*: Are 0.6 M glycerol and 0.3 M sodium chloride isosmotic? Would red blood cells find them both isotonic to the intracellular contents?)

2. What are the colligative properties of a solution?

3. What is the osmotic strength of a 20% sucrose solution? (The molecular weight of sucrose is 342.) How high would the sucrose solution go in a very tall osmometer?

4. Certain types of glomerular disease in the kidney lead to the excretion of plasma proteins in the urine. What connection might this have with the excessive swelling (edema) often observed in various body tissues of patients with these diseases? Can you explain this in terms of principles studied in this exercise?

5. Drowning swimmers take large amounts of water into their lungs. Is it more dangerous to take salt water or fresh water into the lungs? Explain in detail what would happen in the alveoli and capillaries in each instance.

REFERENCES

Capaldi, R. A. 1974. A dynamic model of cell membranes. *Sci. Amer.*, *230*(3):26.
Glynn, I. M., and S. J. D. Karlish. 1975. The sodium pump. *Ann. Rev. Physiol.*, *37*:13.
Hokin, L., and M. Hokin. 1965. The chemistry of cell membranes. *Sci. Amer.*, *313*(3):78.
Rothstein, A. 1968. Membrane phenomena. *Ann. Rev. Physiol.*, *30*:15.

Savitz, D., V. W. Sidel, and A. K. Solomon. 1964. Osmotic properties of human red cells. *J. Gen. Physiol., 48*(1):79.

Schultz, S. G., R. A. Frizzell, and H. N. Nellans. 1974. Ion transportation by mammalian small intestine. *Ann. Rev. Physiol., 36*:51.

Solomon, A. K. 1962. Pumps in the living cell. *Sci. Amer., 207*(1):100.

Weissmann, G., and R. Claiborne (eds.). 1975. *Cell Membranes: Biochemistry, Cell Biology, and Pathology*. Hospital Practice Pub. Co., Inc., New York.

EXERCISE 18

Name _____

Laboratory Section _____ Date _____

RESULTS AND CONCLUSIONS

Use of the Osmometer in Studying Osmotic Phenomena

Osmometer	Osmometer Reading After						
	10 min	20 min	30 min	40 min	50 min	60 min	120 min
1							
2							
3							
4							
5							

Plot graphs showing the rate and direction of flow for each of the osmometers. Does a doubling of the concentration of a solution double its osmotic strength?

221

Living Cells as Osmometers

Erythrocytes

Tube #	NaCl Concentration (%)	Appearance of Supernatant	Tonicity	Microscopic Observations
1	0			
2	0.1			
3	0.2			
4	0.3			
5	0.4			
6	0.5			
7	0.6			
8	0.7			
9	0.8			
10	0.9			
11	1.0			
12	2.0			

What concentration of NaCl is isotonic to the red cell contents?

EXERCISE 19

Enzymes

OBJECTIVES In this exercise we shall study several of the factors governing enzyme activity. Introductory procedures deal with various types of enzymatic processes, after which we consider the effect of substrate concentration and inhibitors on the activity of the catabolic enzyme β-galactosidase. An interesting biomedical film, *PKU—Preventable Mental Retardation,* is used to illustrate the role of enzymatic processes in the human body and to demonstrate the serious consequences of inherited enzyme deficiency.

An extraordinary feature of all living organisms is the extreme versatility they exhibit in carrying out a wide variety of chemical processes. These reactions include the degradation of nutrient materials (catabolism), the synthesis of unique organic molecules (anabolism), and the interconversion of organic substances in carbon and energy metabolism (sometimes called amphibolic reactions). The countless processes that comprise intermediary metabolism occur through the agency of specific biological catalysts called enzymes.

Enzymes are globular proteins of highly specific structure. Ordinarily, a given enzyme catalyzes a single chemical reaction, although some enzymes act on a few different chemically related substrates. Quite obviously, thousands of different enzymes are needed to catalyze the many reactions occurring within the human body. Enzymes speed up reactions that would otherwise occur at very much slower rates. One of the essential features of enzyme action is that it allows reactions to take place under conditions of temperature and acidity compatible with life, rather than under the extreme circumstances of high heat and low pH often required for these reactions in the chemistry laboratory.

An enzyme participates in a reaction by binding chemically with the reacting substrate, but is itself regenerated in a truly catalytic fashion. The basic theory of enzyme action assumes a combination between enzyme, E, and substrate, S, to form an enzyme–substrate complex, ES, in a freely reversible reaction.

$$E + S \rightleftharpoons [ES] \longrightarrow Products + E \tag{1}$$

The subsequent breakdown of ES leads to the formation of products and regeneration of the enzyme for further participation in the reaction. Many enzymes can catalyze the breakdown of tens of thousands of substrate molecules each second.

Mathematical analysis of the effect of substrate concentration on the rate of enzyme action was first developed in 1913 by the German biochemists L. Michaelis and M. L. Menten. The basic equation shows that the rate or velocity, v, of a reaction is related to the substrate concentration, [S], as follows:

$$v = \frac{v_{max}[S]}{K_m + [S]} \qquad (2)$$

v = velocity of the reaction at a given value of [S]
S = substrate concentrtion
v_{max} = maximum velocity the reaction can attain when the enzyme is fully saturated with substrate (a constant for a given set of experimental conditions, e.g., pH, temperature, enzyme concentration)
K_m = substrate concentration at which the actual velocity, v, is one half the maximum velocity, v_{max} (also, a constant for a given set of assay conditions)

Figure 19-1 presents a graphic representation of the Michaelis–Menten equation. You can verify the correspondence between the equation and the curve in the figure by constructing a simple graph for yourself. Notice that the Michaelis–Menten equation may be written in a more familiar fashion as $y = ax/(b + x)$, where a and b are constants. Assume that a (or v_{max}) = 10 (i.e., that the maximum velocity of a certain enzyme is 10) and that b (or K_m) = 3 (i.e., that the enzyme works at half its maximum velocity when the substrate concentration is 3). Calculate a series of rates as [S] is increased from 1 to 20. Plot a graph of velocity versus substrate concentration to obtain a typical rectangular hyperbole. Notice especially that v never fully attains v_{max}, even at very high values of [S]. In what range is the velocity linearly proportional to the substrate concentration? Figure 19-2 is a Lineweaver–Burke plot of the data presented in Figure 19-1. In order to make a Lineweaver–Burke plot, one graphs the

FIGURE 19-1 Velocity of an enzyme reaction as a function of the substrate concentration. v_{max} is the maximum velocity when the enzyme is fully saturated with substrate. K_m is the substrate concentration at which the velocity is one half of v_{max}.

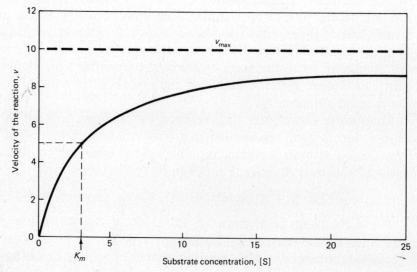

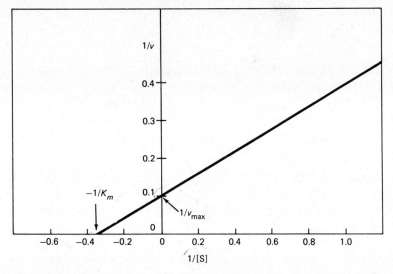

FIGURE 19-2 Lineweaver-Burke plot of the data from Figure 19-1. Abscissa is the reciprocal of the substrate concentration (1/[S]); ordinate is the reciprocal of the velocity (1/v).

reciprocal of the velocity (1/v) versus the reciprocal of the substrate concentration (1/[S]). The resulting straight line intersects the y-axis as $1/v_{max}$ and intersects the x-axis as $-1/K_m$.

The particular region on the enzyme surface where the substrate-binding groups are located is known as the active site. In most instances the active site is a pocket or groove that is complementary in structure to the three-dimensional configuration of the substrate. This relationship is sometimes referred to as the lock-and-key fit. The chemical affinity between enzyme and substrate resides in the specific chemical groups of which both are composed. Hydrophobic interactions, ionic attractions, and hydrogen bonding all participate in binding enzyme and substrate.

Inhibition of enzyme activity can be brought about by substances that are similar but not identical in structure to the normal substrate. These analogs have affinity for the active site and prevent the real substrate from complexing with the catalyst. This form of inhibition is ordinarily termed competitive inhibition. Noncompetitive inhibitors inactivate enzyme function by binding at locations other than the active site.

We shall study several basic types of enzymatic processes, and investigate in detail the specific conditions that govern the rates at which enzymes work. Two alternative procedures are provided. The first series of tests provides a basic introduction to enzyme chemistry for students with minimal preparation in this area. A second series of experiments, with the enzyme β-galactosidase, provides the basis for a more advanced study of the specific conditions that govern the rates at which enzymes work.

TYPES OF ENZYMATIC PROCESSES

The Novo Enzyme Kit forms the basis for a useful introduction to different types of enzymatic reactions, including the following:

(1) Enzymes for cheese production.
(2) Enzymes of digestion: the conversion of protein to amino acids.

228 Section VII] *Cellular and Subcellular Processes*

(3) Enzymes as laundry aids.
(4) Gelatin digestion as means of recovering silver from photographic films and emulsions.
(5) Enzymes of digestion: the conversion of starch to sugar.
(6) Enzymatic desizing of fabrics.
(7) The use of pectic enzyme in the fruit juice industry.

Materials

test tubes and test tube racks
37°C water baths
boiling water baths
1 ml pipet
5 ml pipets
Novo Enzyme Kit (see Note)

Note: Available from Carolina Biological Supply Co., Burlington, NC 27215.

Procedures

1. Detailed procedures for each of the seven reactions listed above are outlined in the Student's Manual provided with the enzyme kit. In each case a purified preparation of the enzyme under study is incubated with an appropriate substrate, and the course of the reaction is assayed by observing the *disappearance* of the substrate and/or the *appearance* of the reaction products(s).
2. Determine which reactions you are going to investigate, obtain the materials you need, and follow the instructions outlined in the Student's Manual.

FACTORS GOVERNING ENZYME REACTIONS

β-Galactosidase, a catabolic enzyme involved in the breakdown of milk sugar (lactose) and other more complex sugars, can be readily studied in a laboratory exercise. The equations below summarize two of the reactions catalyzed by this enzyme. A convenient assay makes use of a nonphysiological substrate, o-nitrophenyl galactoside (o-NPG), to study enzyme activity. Hydrolysis of this substrate by the enzyme yields two products, the monosaccharide galactose and a bright yellow compound, o-nitrophenol. The appearance of yellow color affords a simple measure of the course of the reaction.

$$\text{Lactose} + H_2O \longrightarrow \text{Glucose} + \text{Galactose} \qquad (3)$$

$$\underset{\text{(colorless)}}{o\text{-Nitrophenyl galactoside}} + H_2O \longrightarrow \underset{\text{(yellow)}}{o\text{-Nitrophenol}} + \text{Galactose} \qquad (4)$$

Exercise 19] Enzymes

Materials

- test tubes (20 per two students)
- test tube racks (1 per two students)
- 1 ml pipets (6 per two students)
- 5 ml pipets (4 per two students)
- β-Galactosidase (20 ml per two students; see Note)
- o-nitrophenyl-β-D-galactopyranoside (o-NPG) at 0.01 M concentration (15 ml per two students; see Note)
- 0.7 M sodium phosphate buffer, pH 7.25, 0.7 N Na (40 ml per two students)
- 0.4 M Na$_2$CO$_3$ (100 ml per two students)
- distilled water
- 34°C water bath (optional)
- Spectronic 20 spectrophotometers and tubes, or suitable alternative (1 per two to four students)
- inhibitors of activity (optional): 3.0 M glucose, 0.5 M galactose, 0.1 M HgCl$_2$ (5 ml of each per two students)

Note: The Sigma Chemical Co., P. O. Box 14508, St. Louis, MO 63178 supplies several forms of β-galactosidase (EC No. 3.2.123) that can be used in this exercise. Product #G1875 prepared from bovine liver and product #G8504 from *Escherichia coli* are entirely suitable. The lyophilized bovine enzyme contains its own buffer and may be dissolved in water at a concentration of 0.2 unit/ml. The purified *E. coli* enzyme may be diluted in 0.05 M phosphate buffer (pH 7.25) to a concentration of 0.2 unit/ml.
NPG is also available from Sigma.

Procedures

Standard Assay (5 ml) and Basic Properties of the Enzyme

1. The substrate, o-NPG, is colorless, but the reaction product, o-nitrophenol, has an intense yellow color, especially at alkaline pH.
2. To a series of test tubes add 2.0 ml of 0.7 M phosphate buffer (pH 7.25). To each tube add 1.0 ml of enzyme solution and 1.0 ml of water. Start the reaction by adding 1.0 ml of o-NPG, and incubate the tubes at room temperature (or at 34°C). A control tube without enzyme should also be set up. Yellow color develops in a period of 10–20 minutes. The reaction is stopped by adding 5 ml of 0.4 M Na$_2$CO$_3$, which inhibits enzyme activity and enhances the yellow color of o-nitrophenol.
3. Read the absorbance values in the spectrophotometer at 420 nm. If the absorbance is greater than 0.7 or so, dilute the solution with Na$_2$CO$_3$ and read the value again.
4. The time course of the reaction may be determined by incubating the tubes for various lengths of time. This assay may also be used to study the effect of varying enzyme concentrations on the rate of reaction.

Effect of Substrate Concentration

1. The Michaelis–Menten kinetics for β-galactosidase may be studied using the 5 ml assay. To a series of 6 tubes pipet the following solutions:

Tube #	Buffer	Enzyme	Water	o-NPG	Final Concentration of o-NPG
1	2.0 ml	1.0 ml	1.8 ml	0.2 ml	$0.4 \times 10^{-3}\ M$
2	2.0 ml	1.0 ml	1.6 ml	0.4 ml	$0.8 \times 10^{-3}\ M$
3	2.0 ml	1.0 ml	1.4 ml	0.6 ml	$1.2 \times 10^{-3}\ M$
4	2.0 ml	1.0 ml	1.2 ml	0.8 ml	$1.6 \times 10^{-3}\ M$
5	2.0 ml	1.0 ml	1.0 ml	1.0 ml	$2.0 \times 10^{-3}\ M$
6	2.0 ml	1.0 ml	2.0 ml	0	0

2. Incubate the tubes at room temperature until sufficient yellow color develops (10–30 minutes). Stop the reactions with Na_2CO_3, and read the absorbance values at 420 nm.

Inhibition Studies

1. Use the 5 ml assay to study the effect of various possible inhibitors of enzyme activity such as $HgCl_2$, glucose, and galactose. The final inhibitor concentrations in the assay tubes should be 0.01 M $HgCl_2$, 0.7 M glucose, 0.05 M galactose. Adjust the volume of water in each tube to maintain a total volume of 5 ml.
2. The nature of glucose and/or galactose inhibition (competitive or noncompetitive) may be determined by varying the substrate concentration, as you did above in studying the Michaelis–Menten kinetics, in the presence of 0.7 M glucose. This experiment may be performed simultaneously while studying the effect of substrate concentration by setting up additional tubes containing inhibitor at the specified concentration.

PKU—PREVENTABLE MENTAL RETARDATION

The essential role of enzymes in mediating the various metabolic reactions of the body is dramatically illustrated in those hereditary conditions known as "inborn errors of metabolism." First described in 1908 by the British physician Sir Archibald Garrod, these diseases can be shown in most cases to lie in the absence of one or another of the body's enzymes. Galactosemia, Tay–Sachs disease, albinism, and phenylketonuria (PKU) are well-known examples.

PKU is an autosomally inherited, recessive disease in man, with very serious consequences for the afflicted individual, including severe mental retardation and great loss of muscular control and coordination. The condition can be diagnosed early in infancy by the presence of phenyl ketone in the urine and elevated levels of phenylalanine in the urine and/or blood. Biochemical analysis has shown that the fundamental deficiency resides in the absence of a key enzyme of the liver, phenylalanine parahydroxylase. The reaction sequence in which this enzyme participates is illustrated in Figure 19–3.

When through genetic deficiency the body is unable to synthesize parahydroxylase, the conversion of phenylalanine to tryosine does not take place and phenylalanine accumulates in the blood. Some of the excess is deaminated to a phenyl ketone (phenylpyruvic acid), and this substance has severely toxic effects on the tissues of the brain. Recent evidence suggests that phenyl ketone may interfere with brain metabolism by acting as a competitive inhibitor of a glycolytic enzyme [Bowden and McArthur, 1972]. Fortunately, the condition can be treated with a marked degree of success if it is diagnosed early in infancy.

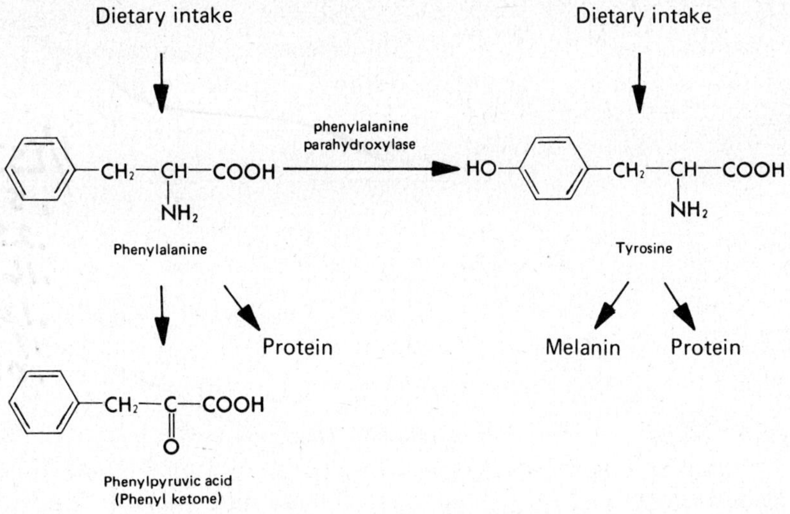

Exercise 19] Enzymes

FIGURE 19-3. Biochemical pathways involved in the metabolism of two aromatic amino acids. PKU patients cannot convert phenylalanine to tyrosine because they lack the enzyme phenylalanine parahydroxylase.

We shall view the excellent medical film *PKU—Preventable Mental Retardation.* This film presents clinical accounts of young patients who have been diagnosed at various ages of infancy and early childhood. Several questions to consider during the viewing are: What are several different ways in which one might treat PKU? Which of these are feasible? Can one truly cure the condition? What are some implications for the population as a whole now that many of the treated individuals can grow to adulthood and become parents?

STUDY QUESTIONS

1. Under suitably controlled laboratory conditions, an investigator measured the activity of the enzyme arginase in the presence of increasing concentrations of its substrate arginine. After 10 minute incubations, the reactions were stopped and the amounts of the product urea were determined by colorimetric analysis. The following data were obtained:

Tube #	$[S]$ [a]	v [b]	$1/[S]$	$1/v$
1	2	0.29	.5	3.45
2	4	0.45	.25	2.22
3	6	0.55	.1667	1.82
4	8	0.62	.125	1.61
5	10	0.67	.1	1.49
6	20	0.80	.05	1.25

[a] In millimoles/liter (mM)
[b] Measured as units of product formed/10 minutes

Plot a graph of v versus $[S]$. Find K_m in millimoles/liter and v_{max} as units of product formed/10 minutes. Use the Lineweaver-Burke method. Indicate K_m and v_{max} on your graph.

*PKU—Preventable Mental Retardation, from Pennsylvania State University (see page 341).

2. The experiments described above were repeated under the same conditions, but in the presence of a certain carefully selected concentration of glutamate ions. The data in this experiment were as follows:

Tube #	[S]^a	v^b
1	2	.17
2	4	.29
3	6	.38
4	8	.44
5	10	.50
6	20	.67

Is glutamate an inhibitor of arginase? If so, is the inhibitor competitive or noncompetitive? Does a Lineweaver–Burke plot supply the answer? How? Do the molecular formulas for arginine and glutamate give any indication of the type of inhibition you might expect?

3. Where is the enzyme β-galactosidase found in the human body? Where is the enzyme lactase found in human body? What are the clinical manifestations of conditions in man in which each of these enzymes is absent? (Consult Holzel et al. [1959], Kretchmer [1972], and Okada and O'Brien [1968].)

4. Phenylketonuria results in the accumulation of the amino acid phenylalanine in the blood. Enzymatic deamination (or transamination) of this substance yields phenylpyruvic acid (PPA), or phenyl ketone as it is commonly called. Bowden and McArthur [1972], suggest a molecular mechanism by which PPA may poison metabolic reactions in the brain. What is the mechanism of inhibition that they propose? Draw the molecular structures of the compounds in question. How would the inhibitory process suggested by Bowden and McArthur alter (1) carbon metabolism in the Krebs cycle; (2) energy metabolism (i.e., the production of ATP); and (3) the synthesis of fatty materials from acetyl CoA? How would this inhibition produce the symptoms shown in the film *PKU—Preventable Mental Retardation*?

REFERENCES

Bowden, J. A., and C. L. McArthur III. 1972. Possible biochemical model for phenyl ketonuria. *Nature,* 235:230.

Holzel, A., V. Schwartz, and K. W. Sutcliffe. 1959. Defective lactose absorption causing malnutrition in infancy. *Lancet,* 1:1126.

Hsia, D. Y-Y. 1966. The diagnosis of carriers of disease-producing genes. *Ann. N. Y. Acad. Sci.,* 134:946.

Jacob, F., and J. Monod. 1961. Genetic regulatory mechanisms in the synthesis of proteins. *J. Mol. Biol.,* 3:318.

Kretchmer, N. 1972. Lactose and lactase. *Sci. Amer.,* 227(4):70.

Okada, S., and J. S. O'Brien. 1968. Generalized gangliosidosis: beta-galactosidase deficiency. *Science,* 160:1002.

Pederson, D. M. 1974. Lecture demonstrations in kinetics relevant to the biology student. *J. Chem. Educ.,* *51*(4):268.

Wallenfels, K. 1962. "β-Galactosidase (Crystalline)." In S.P. Colowick and N. O. Kaplan (eds.), *Methods in Enzymology,* Vol. 5, p. 212. Academic Press, New York.

Name _____

Laboratory Section _____   Date _____

EXERCISE 19

RESULTS AND CONCLUSIONS

Types of Enzymatic Processes

For each of the chemical reactions you have studied, write a balanced chemical equation to indicate the substrate(s), the product(s), and the stoichiometry of the reaction. What is the physiological significance of the reactions you have studied? What is the significance of the particular experimental conditions under which you have incubated your enzymes?

Factors Governing Enzyme Reactions (β-Galactosidase)

Time Course of the Reaction
Plot a graph showing the appearance of product (absorbance at 420 nm) versus time. Is the reaction linear with time? How much of the substrate has been used up at the longest incubation time? Do you think the reaction remains linear until all of the substrate has been consumed? Why?

Exercise 19] Results and Conclusions

Relation of Amount of Enzyme to Rate
What effect does increasing the amount of β-galactosidase have on the rate as measured in the standard assay? Plot a graph to illustrate your results.

Effect of Varying Substrate Concentration
Plot a graph relating rate of reaction, v, versus substrate concentration, [S]. What are the units of each of these parameters?

Record your data in the table below and calculate the values of $1/[S]$ and $1/v$ as indicated.

Tube #	[S]	1/[S]	v	1/v
1	0.4×10^{-3} M			
2	0.8×10^{-3} M			
3	1.2×10^{-3} M			
4	1.6×10^{-3} M			
5	2.0×10^{-3} M	0.5×10^{-3}		

$2.0 \overline{)1.0}$

Exercise 19] Results and Conclusions

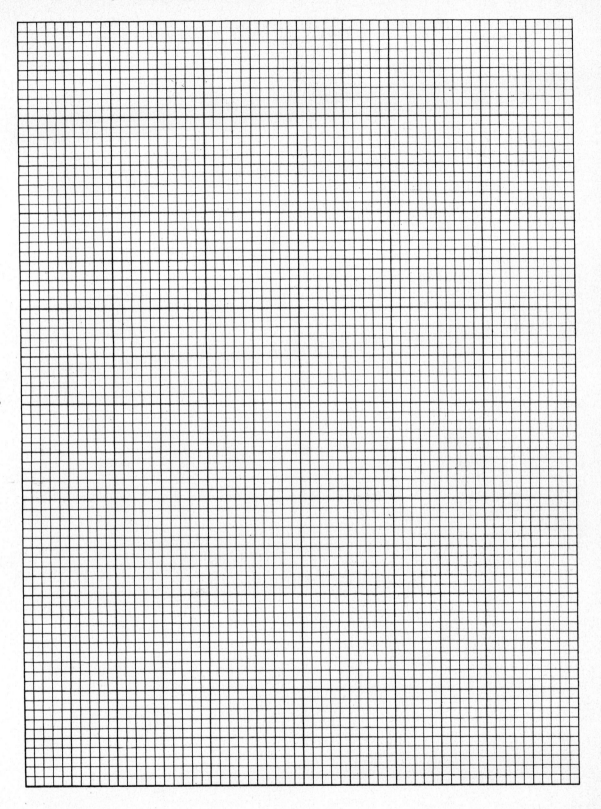

EXERCISE 20

Karyotyping of Human Chromosomes and Amniocentesis

OBJECTIVES In this exercise we shall culture human lymphocytes in suitable nutrient media and prepare metaphase chromosomes for microscopic study; alternatively, we shall examine prepared slides of human chromosomes. Photographs of two metaphase spreads are provided for karyotype analysis. The excellent biomedical film *Prenatal Diagnosis by Amniocentesis* will be shown to illustrate the use of amniocentesis in the prenatal determination of chromosomal defects (Down's syndrome) and enzyme deficiencies (Tay-Sachs disease).

Within the last 20 years, detailed cytological procedures have been established for preparing and studying the chromosomes of man. A commonly employed method involves the growing of certain cell types (leukocytes, skin cells, etc.) in tissue culture for a period of several days, and the subsequent preparation of these cells for microscopic viewing and study.

Lymphocytes, a class of white cell found in the blood, can be induced to divide in a suitable nutrient fluid by the addition of mitotic stimulants such as phytohemeagglutinin (PHA); these cells show a mean generation time of 18-24 hours at 37°C. Over the course of two or three days the lymphocytes divide several times under asynchronous conditions, so that only a small proportion of the cells is in metaphase of mitosis at any given time. After a sufficient period of growth has taken place, the addition of a mitotic inhibitor such as colchicine increases the proportion of cells in metaphase. Subsequent preparation includes washing of the cells by centrifugation, treatment with hypotonic solution (distilled water or diluted serum) to cause the cell nuclei to swell and the bunched chromosomes to spread out, and chemical fixation. Finally the cells are placed on microscope slides and stained, prior to viewing.

Using this technique, cytogeneticists have definitely established the normal diploid chromosome number in man at 46. The photomicrograph provided in Figure 20-1 illustrates a typical metaphase spread and shows the various individual chromosomes. The formal classification of these chromosomes is known as a karyotype. The overall length of the chromatids and the relative position of the centromere along this length are the criteria used in assigning chromosomes to the various groups (A-G).

FIGURE 20-1 [opposite] *Top:* A typical metaphase spread of a human lymphocyte showing 46 chromosomes. *Bottom:* A karyotype prepared from the metaphase spread. Chromosome pairs 1-3 form group A; 4-5 form group B; 6-12, group C; 13-15, group D; 16-18, group E; 19-20, group F; 21-22, group G. Notice the placement of the two sex chromosomes. [*Courtesy of Dr. Kurt Hirschhorn.*]

One of the significant clinical applications of karyotyping has been its use in diagnosing human conditions in which the chromosome number has been altered. The French cytogeneticist J. Lejeune first demonstrated in 1959 that children exhibiting the symptoms of Down's syndrome (mongolism) possess 47 chromosomes instead of the usual 46, of which the 47th is an extra number 21. Since that time many other alterations of chromosome number have been described.

A further clinical application of this work has been in prenatal diagnosis through a process termed *amniocentesis*. Between the twelfth and fourteenth week of pregnancy, fluid is withdrawn from the embryonic sac surrounding the unborn fetus. Tissue culture of living cells found within this amniotic fluid provides sufficient material for karyotype analysis. In an increasingly large number of instances prenatal diagnosis has revealed serious genetic disorders and has been used as the basis for abortion.

Amniocentesis has also been used to diagnose inherited conditions not involving chromosomal aberrations. Recessive mutations, when present in homozygous dosage, lead to enzyme deficiencies that in some cases can be diagnosed through biochemical study of fetal cells grown in tissue culture. Tay-Sachs disease, a very serious inherited condition afflicting the central nervous system, and numerous other conditions can be detected in this manner.

The ethical issues inherent in much of this work are complex, and concern, among other things, a discussion of the child's "right to be normal" as against the "right to life" and the parents' "right to choose." Students interested in pursuing these themes should read Valenti [1968] and the inspiring autobiography of a mongoloid child by Hunt [1967]. In addition, the numerous publications of the Hastings Institute of Society, Ethics and the Life Sciences provide source materials for a thorough study of the ethical concerns arising from recent medical advances.

We shall study the basic features of human chromosomes, karyotyping, and amniocentesis. Two different sets of procedures make the exercise adaptable to a variety of teaching situations.

CHROMOSOME STUDY AND KARYOTYPE ANALYSIS

Materials

prepared slides of human chromosomes
microscopes
scissors
paper paste or rubber cement

Procedures

1. Examine a prepared slide showing human chromosomes. Systematically scan the slide until you locate metaphase chromosomes. Carefully study the chromosomes

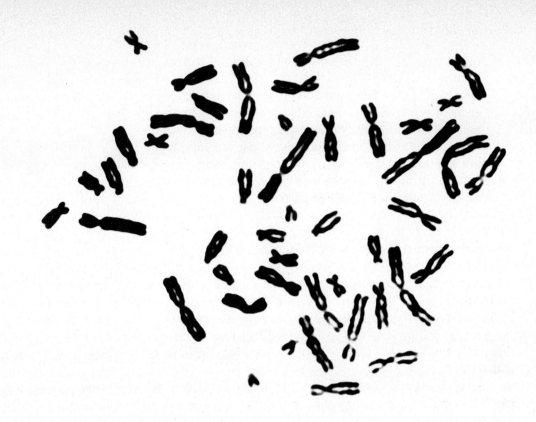
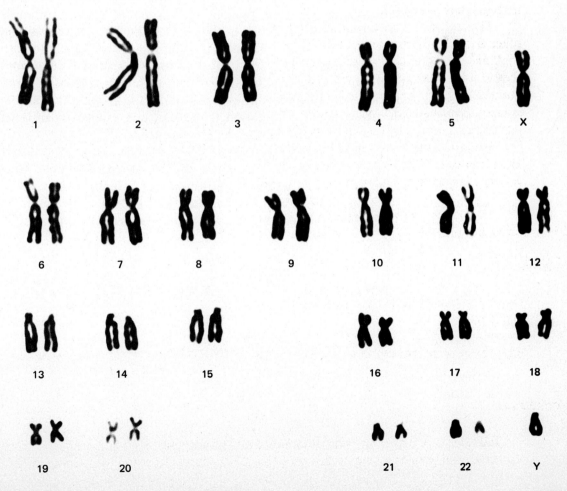

243

under the highest magnification possible with your microscope. Note any unusual features.
2. Remove Figures 20-2 and 20-3 from your laboratory book and carefully cut out each of the chromosomes. Arrange them systematically, according to accepted conventions, on the karyotype analysis sheets provided.
3. What is the sex of the individual whose karyotype you have constructed? What are the general features of the chromosomes in each of the lettered groups? Which chromosome assignments present difficulties or ambiguities? What abnormal conditions are illustrated?

PRENATAL DIAGNOSIS BY AMNIOCENTESIS

In this part of the laboratory period we shall view the excellent biomedical film *Prenatal Diagnosis by Amniocentesis.** Two actual case histories are considered in which prospective parents seek genetic counseling concerning as yet unborn children.

In one instance, a woman has previously given birth to a child with Down's syndrome. Through karyotype analysis it has been discovered that this phenotypically normal woman has only 45 chromosomes, one of which represents a chromosome translocation (15/21 translocation). Such women may pass the 15/21 translocation and a number 21 to a child, who also receives a number 21 from the father and is thereby afflicted with Down's syndrome. The results of amniocentesis are shown in the film and the findings are discussed.

In the other case, two parents who have previously had a child with Tay-Sachs disease seek advice concerning a second pregnancy. Tay-Sachs disease (infantile amaurotic idiocy) is found in high frequency in the Ashkenazic Jews of central and eastern Europe. In New York City, about 1 person in 30 of the Jewish population is a carrier. The disease is fatal and is inherited as an autosomal recessive condition. Affected children appear normal and healthy at birth, but within six months begin to show marked deterioration of brain and nervous function. By the age of one year, the child is afflicted with severe mental retardation, blindness, and extensive paralysis; death ensues within three or four years. There is no known cure.

The karyotype of a Tay-Sachs patient is perfectly normal because the defect resides in a single gene mutation with no obvious changes in chromosomal morphology. Enzyme studies of cultured fetal cells, however, can determine the level of hexoseaminidase A, an enzyme of intermediary metabolism known to be absent in Tay-Sachs patients. In the film, the actual process of amniocentesis is shown, and the parents are informed of the results.

TISSUE CULTURE AND KARYOTYPE ANALYSIS

Materials

70% alcohol
sterile finger lancets
sterile cotton

**Prenatal Diagnosis by Amniocentesis,* from Milner-Fenwick (see page 340).

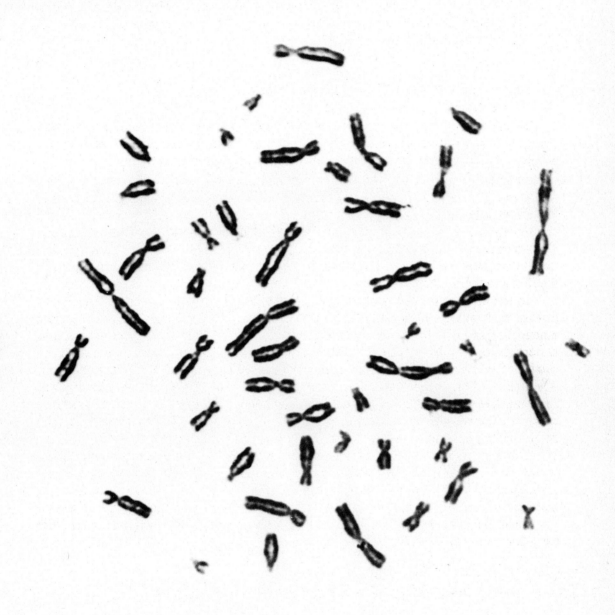

FIGURE 20-2 Metaphase chromosomes for preparation of a karyotype. [*Courtesy of Dr. Kurt Hirschhorn.*]

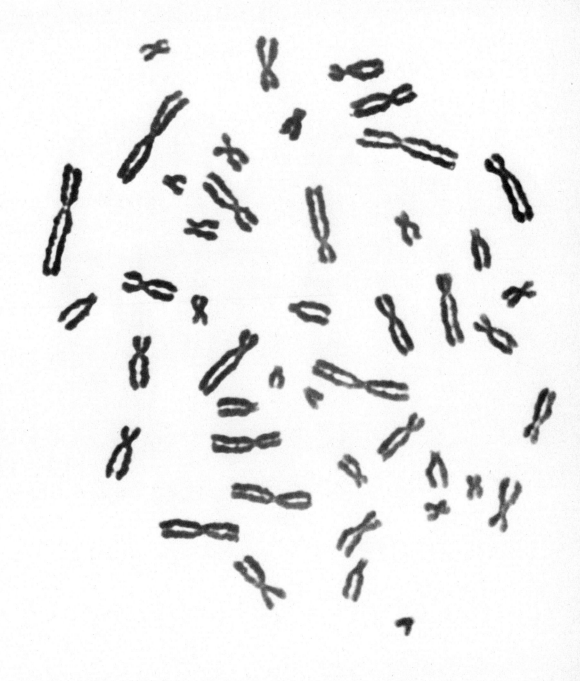

FIGURE 20-3 Metaphase chromosomes for preparation of a karyotype. [*Courtesy of Dr. Kurt Hirschhorn.*]

Exercise 20] Karyotyping of Human Chromosomes and Amniocentesis

chromosome analysis kit including tissue culture growth media and assorted reagents (see Note)
37°C incubator
clinical centrifuges (1 per four to eight students)
centrifuge tubes (6 per two students)
Pasteur pipets fitted with rubber bulbs
hypotonic solution and fixative as specified in the chromosome analysis kit
microscope slides and coverslips
stain solutions (acetoorcein or Giemsa stain)
permount
microscopes

Note: Various highly suitable and convenient media are available from Difco Laboratories, Detroit, MI 48232; Grand Island Biological Co., 3175 Staley Rd., Grand Island, NY 14072; Carolina Biological Supply Co., Burlington, NC 27215; Wards, P. O. Box 11712, Rochester, NY 14063. Detailed instructions for use of various kits are supplied.

Procedures

1. A small sample of blood must be provided for tissue culture. Disinfect your finger with alcohol, allow it to dry and puncture it with a sterile lancet. Collect several drops of blood in the container of growth medium. Remember that this procedure must be performed under sterile conditions, as the nutrients in the growth medium support bacterial growth very amply. Antibiotics are present in the nutrient solution to retard the growth of contaminants. Incubate the growth tubes at 37°C for three days.
2. (It may be necessary for the instructor to perform this step in preparation for the laboratory period.) After 72 hours of growth, add a solution of mitotic inhibitor, as supplied with the kit, to the growth tubes. The cells should be incubated for a further few hours, as directed.
3. Harvest the cells by transferring them with a Pasteur pipet to a clean conical centrifuge tube. Follow the directions provided with the kit as you prepare the cells for microscopic examination.
4. After the cells have been stained, observe them and scan your preparation for metaphase figures. If suitable photographic equipment is available, the chromosomes may be photographed and a karyotype constructed.

STUDY QUESTIONS

1. What is nondisjunction, and how does it lead to alterations of the chromosome number? Is the 21st pair of chromosomes the only pair to undergo nondisjunction? How does the XYY condition that has been described in some men arise? What are some other examples of trisomic conditions?

2. What is the purpose of adding each of the following to the growth medium in which human cells are cultured: antibiotics, phytohemeagglutinin, colchicine, fetal calf serum? How does colchicine actually work?

3. What are some of the inborn errors of metabolism that can be diagnosed prenatally? How is it possible to detect carriers (i.e., heterozygous individuals) through analysis of amniotic fluid? Friedmann [1971] should be helpful.

4. Valenti [1968] quotes the biologist N. J. Berrill as follows:

If a human right exists at all, it is the right to be born with normal body and mind, with the prospect of developing further to fulfillment. If this is to be denied, then life and conscience are mockery and a chance should be made for another throw of the ovarian dice.

By contrast, Douglas Hunt, the father of Nigel, a mongoloid boy, wrote in his introduction to Nigel's autobiography [Hunt, 1967]:

Nigel is not perfect nor is he a genius, though he may be a genius manqué. He used to have a vile temper but his mother cured him of this completely and with the minimum of physical correction. He still finds money a considerable difficulty . . . but he is learning. This, above all, is the message we, his parents, would like to give to our readers. *Mongoloids can learn and go on learning if they are given the encouragement.* They are children like any others, and like any others they are both a blessing and a discipline. The parents of mongoloids have different problems from those of other parents; they miss some of the joys that other parents have, but they also have joys unknown to the parents of normal children. . . . We would not exchange him for the most brilliant child in the world and we have been richly and abundantly rewarded for all that we have tried to do for him.

What is your position on this complex issue?

REFERENCES

Bergsma, D., et al. (eds.). 1973. *Contemporary Genetic Counseling.* National Foundation–March of Dimes. Vol. IX, No. 4, contains five articles on various aspects of genetic counseling. (Available from the Foundation.)

Etzioni, A. 1973. *Genetic Fix.* Macmillan, New York

Friedmann, T. 1971. Prenatal diagnosis of genetic disease. *Sci. Amer.,* 225(5):34.

Hsu, L. Y. F., et al. 1973. Results and pitfalls in prenatal cytogenetic diagnosis. *J. Med. Genetics, 10*(June 1973):112.

Hunt, N. 1967. *The World of Nigel Hunt: The Diary of a Mongoloid Youth.* Garrett Publications, New York.

Institute of Society, Ethics and the Life Sciences, Hastings, New York. (Regular publications from this organization deal with various topics in biomedical ethics.)

Levitan, M., and A. Montagu. 1971. *Textbook of Human Genetics.* Oxford University Press, New York.

Lipkin, M., Jr., and P. T. Rowley (eds.). 1976. *Genetic Responsibility, On Choosing Our Children's Genes.* Plenum, New York.

McKusick, V. A., and R. Claiborne (eds.). 1973. *Medical Genetics.* Hospital Practice Publishing Co., New York. (A collection of 27 articles covering a variety of topics, ranging from chromosome abnormalities to metabolic disorders to genetic screening and counseling.)

Paoletti, R. A. (ed.). 1972. *Selected Readings: Genetic Engineering and Bioethics.* MSS Information Corp., New York.

Ramsey, P. 1970. *Fabricated Man: The Ethics of Genetic Control.* Yale University Press, New Haven.

Valenti, C. 1968. His right to be normal. *Sat. Rev. Lit.,* Dec. 7, 1968, p. 75.

Name _____

Laboratory Section _____   Date _____

EXERCISE 20

RESULTS AND CONCLUSIONS

What condition(s) are illustrated in each of the karyotypes you have prepared? What is the sex of the individuals under consideration?

Propose genetic mechanisms to account for the mode of inheritance in each of these cases.

254 *Exercise 20] Results and Conclusions*

Karyotype analysis sheet.

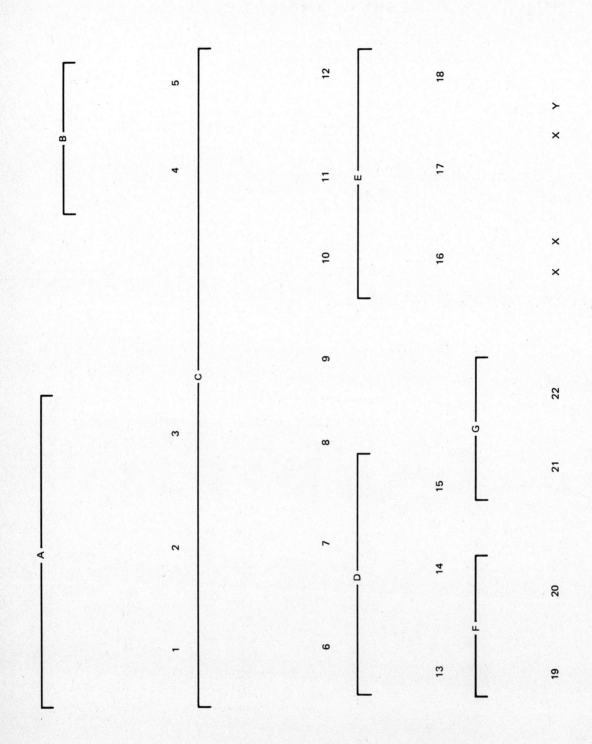

Exercise 20] **Results and Conclusions**

Karyotype analysis sheet.

SECTION VIII

Renal Function

EXERCISE 21

Elementary Urinalysis

OBJECTIVES In this exercise we shall examine several basic properties of a sample of our own urine: (1) color, (2) pH, (3) specific gravity and the concentration of dissolved solids, (4) glucose content, (5) protein content, and (6) possible presence of bacteria. In addition, we shall examine the sediment from a centrifuged sample with the microscope for the presence of solid elements in the urine.

The average adult human being produces about 1-1½ liters of urine each day. This volume represents only a very small portion of the total volume of filtered blood plasma that is processed by the kidneys; as much as 150 liters of fluid may be filtered and reabsorbed daily. Tubular reabsorption of water, therefore, involves the retention of over 99% of the fluid entering the kidney tubules.

Urine itself is a complex solution of metabolic waste products and usually contains about 50-60 grams of dissolved materials per liter. Roughly half of these materials are inorganic substances such as phosphates, sulfates, nitrates, sodium chloride, and various other inorganic ions (calcium, potassium, and magnesium).

Organic substances include urea, uric acid, and creatinine. Amino acids and glucose are ordinarily not found in urine. Urea accounts for over 90% of the nitrogenous material in urine. The principal site of urea production is the liver, where amino groups derived from amino acids by deamination are combined enzymatically with carbon dioxide. Uric acid, a purine compound related to adenine and guanine, is also present in human urine and is a by-product of nucleic acid metabolism. Its concentration in human urine is rather low. By contrast, uric acid is the principal nitrogenous component in the urine of birds and reptiles. Small amounts of creatinine are also excreted by man. Only very small quantities of ammonia are found in either human plasma or urine, as this compound is extremely toxic.

The analysis of urine gives important information regarding the role of the kidney as an organ of excretion and as an organ of homeostasis. In addition, urinalysis is an extremely important diagnostic tool in detecting many abnormal conditions in the human body.

The basic anatomy of the human male kidneys and urinary system is illustrated in Figure 21-1. Study the drawing and identify the following gross anatomical features: abdominal aorta, inferior vena cava, renal artery and vein, renal pelvis, ureter, ureteral orifice, urinary bladder, ejaculatory orifice, urethra, and the prostate gland.

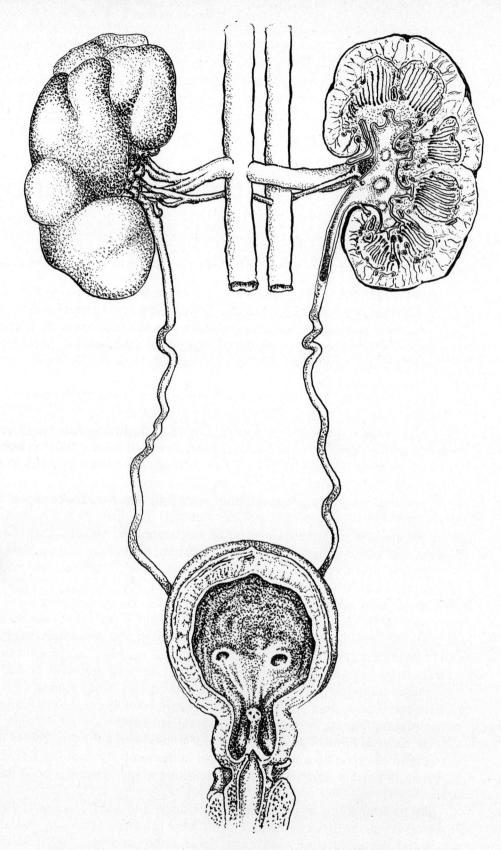

FIGURE 21-1 The urinary system of man. Identify and label the various structures as indicated in the text.

Exercise 21] Elementary Urinalysis

Identify the following structures in the cut-away drawing of the left kidney: renal capsule, cortex, medulla, renal pyramids, renal columns, major calyces, minor calyces, renal papilla. How are the renal tubules arranged in the kidney?

A 24 hour urine collection is often used for quantitative testing. For our purposes a so-called random urine sample is quite satisfactory. Collect a sample in a clean specimen jar, and store the container in the refrigerator if the tests are not to be performed immediately. Carry out the following tests as directed.

COLOR

The color of urine is ordinarily amber yellow due to the presence of urochrome, a breakdown product of hematin, the nonprotein portion of hemoglobin.

Hemoglobin ⟶ Hematin ⟶ Bilirubin ⟶ Urochromogen ⟶ Urochrome

Under certain abnormal conditions, urine may take on a variety of other colors. Nearly colorless urine may be due to extreme dilution of the pigments ordinarily present because of increased water elimination. In diabetes insipidus, a condition in which water reabsorption in the distal convoluted tubules and collecting ducts is markedly reduced, an individual may produce 5-15 liters of highly diluted urine per day.

Milky coloration may be due to the presence of fat globules, pus, or bacterial organisms.

Reddish urine indicates the presence of heme pigments, such as uroporphyrin, uroerythrin, or hemoglobin itself. Such components result from internal hemorrhage, hemoglobinuria, congenital porphyria (a rare inherited disease in which the urine is wine red), or traumatic shock.

Greenish yellow urine may be caused by the presence of bile pigments and is known to occur in jaundice.

Other colors can occur and are generally indicative of metabolic imbalances, various pathologies, or unusual dietary intake.

Procedures

1. Examine your own urine for color and odor. Record your observations on the data sheet.

HYDROGEN ION CONCENTRATION (pH)

The hydrogen ion concentration of fresh urine is highly variable. Usually it is around pH 6, but it may fluctuate anywhere between 4.8 and 8.0. The nature of one's diet has a great deal to do with the acidity or alkalinity of the urine.

Materials

pH paper
pH meter (if available)

Procedures

1. Dip the pH paper into the urine and record the value. A more precise determination can be made if a pH meter is available.

SPECIFIC GRAVITY

The specific gravity of urine is a direct measure of the concentration of dissolved solid materials and is dependent upon dietary intake, fluid intake and elimination, the status of one's metabolism, and environmental conditions. The specific gravity normally varies between 1.010 and 1.030, but it may attain higher or lower values in certaitin pathological conditions.

Procedures

1. Pour approximately 60 ml of fresh urine into the urinometer cylinder and record the temperature of the sample.
2. Float the urinometer in the urine, as shown in Figure 21-2, and read the specific gravity directly from the instrument (at the bottom of the meniscus).
3. Record the value to three decimal places (e.g., 1.018). It is necessary to apply a correction if the temperature of the urine differs from the temperature at which

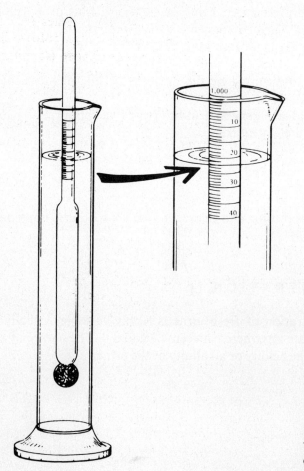

FIGURE 21-2 A urinometer (hydrometer) for determination of the specific gravity of urine.

Exercise 21] Elementary Urinalysis 263

the urinometer was calibrated. For every 3°C below the calibration temperature, subtract 0.001 from the measured specific gravity; for every 3°C above the calibration temperature, add 0.001.

4. Estimate the grams of solids in your urine sample from the specific gravity by using Long's coefficient. Solids in grams per liter are computed by multiplying the last two decimal places of the specific gravity by the factor 2.66.

Example: A 24 hour urine sample had a total volume of 1.20 liters and a specific gravity of 1.015.

$$15 \times 2.66 = 39.9 \text{ grams of solids/liter}$$
$$39.9 \times 1200 = 47.9 \text{ grams of solids/1.2 liters/24 hours}$$

Record the value in your urinalysis report.

GLUCOSE

Under ordinary circumstances very little glucose is found in the urine. A normal adult may excrete a total amount of only 100 mg/day. Concentrations of sugar in normal blood serum, however, vary from 65 to 140 mg/100 ml, depending on how recently the individual has eaten.

The obvious explanation for the absence of glucose in urine is that virtually all of the sugar present in the glomerular filtrate is reabsorbed by the renal tubules. The *renal threshold* for glucose reabsorption is approximately 180 mg/100 ml. In diabetes mellitus blood sugar levels rise to well above the renal threshold values (300–500 mg/100 ml or higher) because of the inability of the body to absorb sugar into the liver and muscle tissue. Under these conditions, large quantities of glucose (as much as several hundred grams a day) are excreted in the urine. Measurement of urinary glucose, therefore, is an important diagnostic measure ordinarily performed as part of routine medical examination.

We will perform a semiquantitative determination of urine sugar using Benedict's reagent. This test is based on the fact that glucose is a reducing sugar; in alkaline solution cupric ions (Cu^{2+}) are reduced to cuprous (Cu^+) as glucose is oxidized.

Materials

Benedict's qualitative sugar reagent (50 ml per two students; see Note)
Erlenmeyer flasks (1 per two students)
boiling water baths and test tube racks to fit baths (1 per six students)
boiling chips
test tubes (10 per two students)
1 ml pipets (3 per two students)
5 ml pipets (1 per two students)
2% standard glucose solution (3 ml per two students)

Note: Benedict's qualitative reagent contains 17.3 grams of $CuSO_4$ (anhydrous), 173 grams of sodium citrate, and 100 grams of Na_2CO_3 in 1 liter of water. The two sodium salts are dissolved in about 800 ml of water with gentle heating; $CuSO_4$ is dissolved separately in 100 ml of water (this reaction evolves heat); and the two solutions are mixed together slowly with constant stirring. The final volume is brought to 1 liter. Clinitest tablets may be substituted for the Benedict's sugar reagent. In this case only test tubes, test tube racks, and pipets are needed.

Procedures

1. Pipet the following solutions into a series of test tubes.

Tube #	Benedict's Reagent	Urine	Water	Glucose Standard
1	5.0 ml	0	0.4 ml	0
2	5.0 ml	0.4 ml	0	0
3	5.0 ml	0	0	0.40 ml
4	5.0 ml	0	0.10 ml	0.30 ml
5	5.0 ml	0	0.20 ml	0.20 ml
6	5.0 ml	0	0.30 ml	0.10 ml
7	5.0 ml	0	0.35 ml	0.05 ml

2. Place all of the tubes into a boiling water bath for 5 minutes. At the end of this time remove the tubes and allow them to cool at room temperature.
3. Mix the contents thoroughly and compare your urine sample (tube #2) with the standards.
4. Estimate the concentration of glucose in your urine. The extent to which this reaction occurs is a measure of the glucose concentration.

PROTEIN

The presence of protein in the urine is a strong indication of kidney dysfunction. Normally, large molecular weight substances such as proteins do not pass across the glomerular membrane into the renal tubules. In acute glomerular nephritis, an inflammatory disease of the kidney, however, many of the glomeruli exhibit greatly increased permeability, allowing protein to leak into the glomerular filtrate. In extreme cases of nephritis, red blood cells appear in the urine.

Materials

test tubes (2 per two students)
test tube racks (1 per two students)
Bunsen burners (1 per two students)
test tube holders (1 per two students)
2% acetic acid (1 dropping bottle per four students)
concentrated nitric acid (5 ml per two students)
5 ml pipets (2 per two students)
rubber propipetting bulbs (several per laboratory section)

Exercise 21] Elementary Urinalysis

Procedures

Heller's Test
1. Using a rubber pipetting bulb, carefully add 5 ml of concentrated nitric acid to a test tube.
2. Incline the tube and pour 5 ml of urine down the side of the tube so that the urine forms a layer above the acid. The formation of a white precipitate at the interface indicates the presence of protein.

Heat Coagulation Test
1. Bring 5 ml of urine to a boil in a test tube. The formation of white precipitate is due to protein or to insoluble phosphates.
2. Add 5 drops of 2% acetic acid. Phosphates dissolve in the acidified solution, while the heat-coagulable proteins remain as a precipitate. Do not add excess acid, as the proteins may also dissolve.
3. Record your results.

BACTERIOLOGICAL EXAMINATION

An estimate of the number of viable bacterial organisms in urine can be made by suitable dilution and plating techniques. Under normal circumstances urine is aseptic, but in instances of urinary tract infection, bacterial counts may exceed 100,000 organisms/ml. Counts of 10,000/ml or less are probably caused by contamination.

Materials

trypticase soy agar Petri plates (2 per two students)
0.1 ml sterile pipets (1 per two students)
1.0 ml sterile pipets (2 per two students)
sterile capped dilution tubes containing 9.0 ml of nutrient broth (or saline) (2 per two students)
sterile capped specimen jars for urine collection (1 per two students)
bent glass rods for spreading bacteria (1 per two students)
small bottles of 95% ethyl alcohol (1 per two students)
Bunsen burners (1 per two students)
wax marking pencils

Procedures

1. Collect a fresh urine sample in the sterilized jar which has been provided.
2. Pipet 1.0 ml of the sample into 9.0 ml of sterile saline. Mix thoroughly. This represents a 10-fold dilution of the original sample. Pipet 1.0 ml of the diluted sample into 9.0 ml of sterile saline and mix. This represents a 100-fold dilution.
3. Using sterile 0.1 ml pipets, plate 0.1 ml sample of the two different dilutions onto the agar dishes. Sterilize a bent glass rod by dipping it in 95% alcohol and passing it quickly through a Bunsen flame. Use the rod to spread the drops onto the agar media. Label the plates, and incubate them upside-down at 37°C for 24 hours.

FIGURE 21-3 Crystalline (*above*) and cellular (*opposite*) elements found in human urine.

CRYSTALS

1. Ammonium magnesium phosphate
2. Calcium oxalate
3. Calcium phosphate
4. Uric acid
5. Calcium carbonate
6. Cystine
7. Cholesterol
8. Hippuric acid
9. Leucine
10. Tyrosine

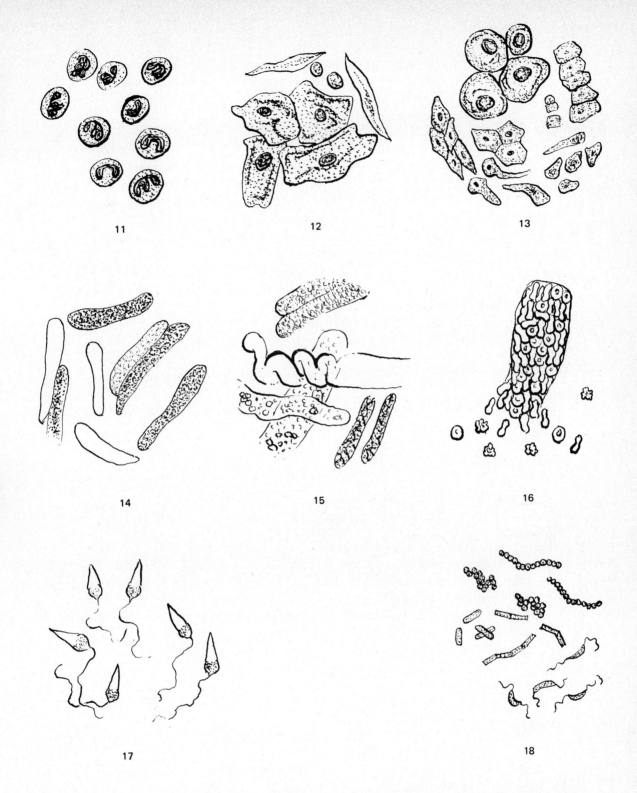

CELLS AND CASTS

11. Leukocytes (pus)
12. Squamous epithelial cells
13. Transitional epithelial cells
14. Hyaline casts
15. Granular casts
16. Blood casts
17. Spermatozoa
18. Bacterial cells

4. Count the number of colonies on each plate, and calculate the bacterial count in the original urine sample, taking into account the various dilutions.

MICROSCOPIC EXAMINATION

Suspended insoluble materials are often present in the urine, many of which may be identified microscopically in a centrifuged urine sample. Many of these are normal excretory products; others are indicators of various pathological states. The presence in the urine of leukocytes and other components of pus is symptomatic of infection in the kidneys or urinary tract. Red blood cells or red cell casts also indicate an underlying pathological condition. (However, the urine sample of a female subject may be contaminated with vaginal blood.) Other solid elements in the urine may include casts (tiny clumps of hardened materials that have taken the form of renal tubules); solidified salts or calculi; and a variety of crystallized materials such as urates, oxalates, phosphates, and certain amino acids.

Materials

clinical centrifuges and centrifuge tubes
10 ml pipets (1 per two students)
microscopes
microscope slides and coverslips
Sedi-Stain

Procedures

1. Centrifuge a urine sample for 10 minutes in a clinical centrifuge. Carefully pour off the clear supernatant fluid and leave the sediment at the bottom of the tube for microscopic examination.
2. Place 1 drop of the sediment on a microscope slide, stain it with 1 drop of Sedi-Stain, and examine the slide under low and high power. Try to identify any elements you find by referring to the drawings in Figure 21-3.

STUDY QUESTIONS

1. How would a high-protein diet affect the pH of the urine? A high-vegetable diet?

2. Under what circumstances would a nondiabetic individual expect to find glucose in his urine as so-called temporary glycosuria?

3. Various elements are sometimes found in urine, such as hemoglobin, bilirubin, the Bence-Jones protein, phenyl ketone, and the amino acid cystine. Cite a disease in which each would be present in the urine.

4. Ketones (such as acetone) are sometimes present in the urine, especially in patients with diabetes mellitus. What is the metabolic origin of the ketone?

5. How does the hemodialysis apparatus work? Make a sketch of the instrument and discuss its operation in detail. To which blood vessels of the body is it attached?

REFERENCES

Davidow, B., et al. 1966. A thin layer chromatographic screening test for the detection of users of morphine and heroin. *Am. J. Clin. Path., 46*:58.

Fleischner, G., and I. M. Arias. 1970. Recent advances in bilirubin formation, transport, metabolism and excretion. *Am. J. Med., 49*:576.

Flocks, R. H. 1970. Urinary tract infection. *Med. Clin. North Am., 54*:397.

Guthrie, R., and A. Susi. 1963. A simple phenylalanine method for detecting phenylketonuria in large populations of newborn infants. *Pediatrics., 32*:338.

Hobbs, J. R. 1966. The detection of Bence-Jones proteins. *Biochem. J., 99*:15P.

King, S. E. 1957. Postural adjustments and protein excretion by the kidneys in renal diseases. *Ann. Intern. Med., 46*:360.

Lassiter, W. E. 1975. Kidney. *Ann. Rev. Physiol., 37*:371.

Merrill, J. P. 1961. The artificial kidney. *Sci. Amer., 205*(1):56.

Smith, H. W. 1953. The kidney. *Sci. Amer., 188*(1):40.

Smith, H. W. 1956. *Principles of Renal Physiology.* Oxford University Press, New York.

EXERCISE 21

Name _____

Laboratory Section _____   Date _____

RESULTS AND CONCLUSIONS

Characteristics of Urine

       Color_____

       pH_____

       Specific gravity_____

       Concentration of dissolved solids_____grams/liter

       Glucose concentration_____

       Protein concentration_____

       Bacterial count_____

Microscopic Examination

In the space below draw any of the solid elements you have located in your sample and provide a tentative identification.

EXERCISE 22

Regulation of Water and Salt Balance by the Kidneys

OBJECTIVES In this exercise we shall examine the role played by the kidneys when presented with excessive water and/or sodium chloride and sodium bicarbonate.

Regulation of the composition and volume of the blood depends largely upon proper functioning of the kidneys. When the normal constituents of the blood, such as salts, glucose, and amino acids, rise above certain concentrations, they are excreted. When excess acid is produced in metabolism, the urine turns acid; when bases are present in excess, the urine is alkaline. Even the total volume of blood is under careful control. When the blood volume tends to increase, as after the consumption of large quantities of water, the urine is copious and dilute; when the blood volume is low, the flow of urine decreases.

Urine is derived from blood. Essentially everything that is present in urine has first been present in the blood, from which it has been separated by the kidneys. In the resting state, a considerable amount of blood plasma is delivered to the kidneys every minute. This volume, which may approximate 1.2 liters/minute in an average-sized person, constitutes as much as one quarter of the total cardiac output. Of the 1.2 liters, about 125 ml is filtered into the glomeruli, but only 1–2 ml eventually appears as urine. Obviously, most (over 99%) of the fluid is reabsorbed from the kidney back into the blood, largely in the proximal convoluted tubules. Fine control over the reabsorption of water in the kidney is brought about by ADH, the antidiuretic hormone of the neurohypophysis.

Salts, amino acids, glucose, and many other substances diffuse freely into the glomerular capsules, but are reabsorbed into the blood as the fluid in which they are dissolved passes along the renal tubules. Certain substances in the blood do not appear in the urine until their concentration exceeds a certain critical level, the threshold value, for that substance. For example, the threshold level for glucose is approximately 180 mg/100 ml of blood (with rather large individual variation); this finding explains why a normal person has virtually no sugar in his urine, even though the substance is present in abundance in his blood and glomerular filtrate. Sodium chloride is actively reabsorbed from the tubules, but normal urine usually contains some salt. Regulation of salt reabsorption is effected, in part, by aldosterone, a hormone of the adrenal cortex. Even relatively small changes in the vital reabsorptive function may result in profound changes in the output and composition of the urine, and in the overall composition of the blood and other body fluids.

Materials

tap water for drinking (approx. 750 ml per drinker)
0.9% NaCl solution for drinking (approx. 750 ml per drinker)
5.0% NaCl solution for drinking *or* coated salt tablets (approx. 75 ml per drinker)
0.1% sodium bicarbonate for drinking (approx. 400 ml per drinker)
drinking glasses
specimen collecting jars (6 per participating student)
100 or 250 ml graduated cylinders (1 per group)
urinometers and urinometer cylinders (1 per group)
pH paper
25 × 250 mm test tubes (10 per group)
test tube racks (1 per group)
10 ml burets and ringstands with clamps (1 per group)
1.0 ml pipets (8 per group)
standard salt solutions for titration: 1%, 2%, 3% (50 ml of each per laboratory section)
2.9% (0.171 M) $AgNO_3$ (100 ml per group)
20% potassium chromate (3 ml per group)

Procedures

1. The class will be divided into five groups for this experiment.
   Any student with a history of heart or kidney trouble of any kind or one who is on a low salt diet for medical purposes should not participate.
   a. Members of group I will drink tap water, 10 ml/kg of body weight; for example, an 80 kg (176 lb) student will drink 800 ml of water.
   b. Group II will drink 0.9% NaCl solution, 10 ml/kg of body weight; a 70 kg (154 lb) student will drink 700 ml of solution.
   c. Group III will drink 5.0% NaCl solution, 1.0 ml/kg of body weight; a 75 kg (165 lb) student will drink 75 ml of solution. This solution is rather unpleasant to drink, and students in Group III may find it easier to ingest coated salt tablets to give a corresponding amount of salt. **(Be absolutely certain to take the correct amount of salt.)** A little water may be drunk to assist in swallowing these tablets. If any student in group III experiences nausea or cramps, he should not continue.
   d. Group IV will drink 0.1% sodium bicarbonate, 5 ml/kg of body weight; a 70 kg (154 lb) student will drink 350 ml of bicarbonate.
   e. Group V will drink nothing and serve as a control for the remainder of the class.
2. All participants must empty their bladders shortly before the start of the experiment. This sample is retained as a control. At time 0, each group drinks the appropriate solution.
3. Collect urine samples every 30 minutes (if possible) for 2½ hours and analyze them as described below. If you are unable to produce a sample at a given time, hold off until the next interval.
4. Measure the volume of each sample with a graduated cylinder and note the color, odor, and general appearance. Check the pH of each sample with pH paper. Record all your observations.

Exercise 22] Regulation of Water and Salt Balance by the Kidneys 275

5. Determine the specific gravity of each sample with a urinometer and calculate the concentration of solid materials as described in Exercise 21.
6. Determine the salt concentration (grams/100 ml and grams/sample) by titration of a 1.0 ml sample. One would like to have a simple quantitative assay for sodium ions, but the use of a flame photometer or a sodium electrode is beyond this course. Accordingly we shall measure the chloride concentration by titrating urine with $AgNO_3$, and compute the total salt concentrations from the data obtained. The titration is carried out as follows:

   a. Fill a 10 ml buret to the 0 mark with 2.9% $AgNO_3$.
   b. Pipet a 1.0 ml urine sample into a 25 × 150 mm test tube, add 1 drop of potassium chromate, and very carefully titrate this sample with $AgNO_3$.
   c. AgCl will appear immediately as a white precipitate. The endpoint occurs when all the chloride has been precipitated and silver chromate (brown) first appears.

It is strongly recommended that students perform practice titrations with standard salt solutions. What volume of $AgNO_3$ is needed to titrate 1.0 ml of 1% salt solution? 1.0 ml of 2% salt solution? 1.0 ml of 3% salt solution? How can one use these values to calculate the salt concentration of the urine samples?

STUDY QUESTIONS

1. Explain in detail how antidiuretic hormone (ADH) and aldosterone regulate water balance and salt balance in the body. What is the actual experimental evidence underlying your contentions?

2. Diabetes insipidus is an inherited disease (sex-linked recessive) in man and exists in two distinct physiological forms. Two separate genetic loci (both sex-linked) are known to produce the typical phenotype in which large volumes (5-15 liters) of highly dilute urine are excreted each day. What is the molecular basis of each of these two forms of the disease? What therapeutic measures are used?

3. How does the kidney regulate the pH of the blood? Cite specific mechanisms by which acid/base imbalances are controlled.

REFERENCES

Bricker, N. S. 1970. The physiologic basis of sodium excretion and diuresis. *Adv. Intern. Med., 16*:17.

Elkinton, J. R., and T. S. Danowsky. 1955. *The Body Fluids: Basic Physiology and Practical Therapeutics.* Williams and Wilkins, Baltimore.

Grantham, J. J. 1976. Fluid secretion in the nephron: relation to renal failure. *Physiol. Rev., 56*:248.

Latham, W. 1968. The renal excretion of hemoglobin: regulatory mechanisms, and the differential excretion of free and protein-bound hemoglobin. *J. Clin. Invest., 38*:652.

Morel, F., and C. de Rouffignac. 1973. Kidney. *Ann. Rev. Physiol., 35*:17.

Orloff, J., and M. Burg. 1971. Kidney. *Ann. Rev. Physiol., 33*:83.

Scholander, R. F. 1957. The wonderful net. *Sci. Amer., 196*(4):96.

Schrier, R. W., and H. E. de Wardener. 1971. Tubular reabsorption of sodium ion. *N. Eng. J. Physiol., 285*:1731.

Share, L., and J. R. Claybough. 1972. Regulation of body fluids. *Ann. Rev. Physiol., 34*:235.

Smith, H. W. 1956. *Principles of Renal Physiology*. Oxford University Press, New York.

Windhager, E. E. 1969. Kidney, water and electrolytes. *Ann. Rev. Physiol., 31*:198.

EXERCISE 22

Name _____

Laboratory Section _____   Date _____

RESULTS AND CONCLUSIONS

Analysis of Urine Samples

Group # _____

Body weight _____ lb _____ kg

Solution drunk _____   Volume _____

Sample	Volume	Specific Gravity	Solids (grams/100 ml)	pH	NaCl Concentration (grams/100 ml)	NaCl (grams/sample)	Color, Appearance
Control							
30 min							
1 hr							
1½ hr							
2 hr							
2½ hr							

Total volume of urine collected _____ ml

Total amount of salt excreted _____ grams

Collect and tabulate the class data for each of the several groups participating in this exercise.

Calculate the volume of urine and amount of salt excreted for each individual (or for members of a group averaged) as volume or amount excreted per kilogram body weight. Plot separate graphs showing milliliters of urine excreted per minute per kilogram of body weight versus time, specific gravity versus time, salt concentration versus time. How can you account for the results you have obtained?

What effect did the bicarbonate solution have on the various parameters measured in this experiment? How can you account for these results?

Exercise 22] Results and Conclusions

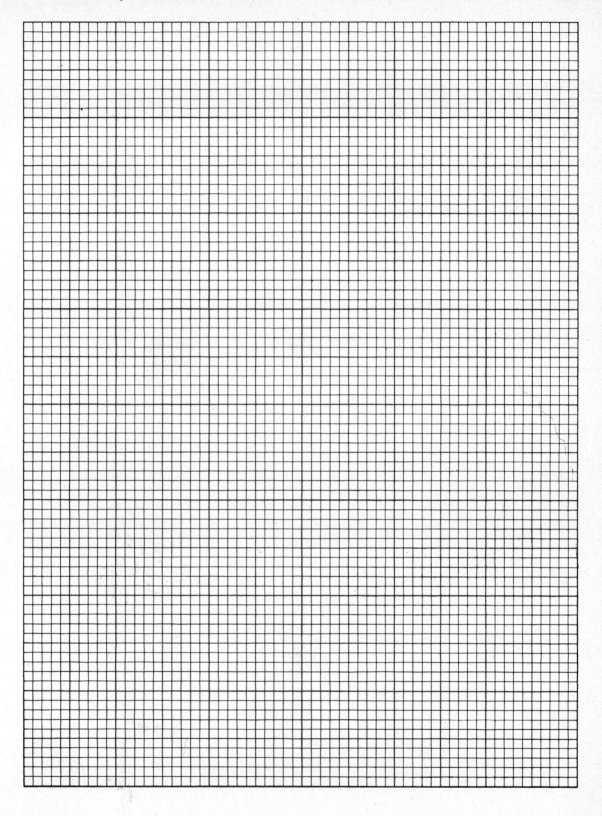

$$60 \overline{)3750} 62$$
$$360$$
$$150$$

120 ml

$120 \overline{)3750}$

EXERCISE 23

Renal Clearance— A Test of Kidney Function

OBJECTIVES In this exercise we shall become familiar with the concept of renal clearance as a clinical measure of kidney activity. We shall test the renal clearance of urea on ourselves.

The chief role of the kidney as an excretory organ is to remove waste products from the blood. As blood plasma filters across the glomerular membranes and passes along the kidney tubules, over 99% of the fluid is reabsorbed back into the blood. Materials such as urea and other waste products, however, are left behind in the tubules, and the reabsorbed fluid is said to be "cleared" of these wastes. The normally functioning human kidney clears urea from the equivalent of 60-70 ml of plasma each minute. The concept of clearance, then, is a measure of renal activity and is often used in the clinical evaluation of overall kidney functioning.

Renal clearance is sometimes measured by injecting a known amount of inulin, a complex polysaccharide isolated from artichoke plants, into the bloodstream. After a certain period of time a urine specimen is collected and the amount of inulin in the sample is assayed. Knowing the concentrations of inulin in the blood and urine and the volume of urine produced in the time interval, one can calculate the renal clearance value as follows:

$$\text{Clearance (ml plasma cleared/minute)} = \frac{\text{mg excreted in urine/minute}}{\text{mg inulin/ml of blood plasma}}$$

As an illustrative example consider the following case: A sterile solution of inulin was injected into the bloodstream of an individual to give a final inulin concentration of 50 mg/100 ml of blood. After 1 hour the individual voided 120 ml of urine (all of which was known to have been produced by the kidneys during that 1 hour), and this sample was shown by analysis to contain 3750 mg of inulin. What is the renal clearance value in this case? (See if you can solve the problem before consulting the solution below.)

Answer:

$$\text{Clearance} = \frac{62.5 \text{ mg inulin excreted/minute}}{0.5 \text{ mg inulin/ml blood plasma}}$$

$$= 125 \text{ ml blood plasma cleared/minute}$$

A value of 125 is typical for renal clearance of inulin; a reading significantly less than this is usually taken as a sign of kidney malfunction.

Renal clearance of urea can be computed if the following quantities are known.

(1) Concentration of urea in the blood plasma (mg urea/ml plasma).
(2) Concentration of urea in the urine (mg urea/ml urine).
(3) Rate of urine formation (ml/minute).

We shall measure these quantities and use them to calculate the clearance value for urea.

At the beginning of the laboratory period, empty your bladder and discard the sample. Note the time exactly. Approximately 1 hour later, empty your bladder again and save the sample, which will be used for the urea determination and to measure the urine flow per minute over the 1 hour time interval. An alternative is to bring a 12 or 24 hour urine sample to the laboratory. Since urine is susceptible to loss of urea through bacterial action, the addition of thymol is recommended in addition to refrigeration. Urine production, expressed as ml/minute, is calculated for the time period over which the sample was collected.

ASSAY OF UREA CONCENTRATION USING COMMERCIAL UREASE

Urease, an enzyme originally isolated from jack-bean meal, catalyzes the hydrolysis of urea according to the equation

$$NH_2-\overset{O}{\underset{\|}{C}}-NH_2 + H_2O \xrightarrow{urease} 2\,NH_3 + CO_2$$

The reaction can be followed by assaying the ammonia produced in this conversion. The assay kit provides reagents for the quantitative estimation of ammonia.

Materials

0.1 ml pipets (2 per two students)
1 ml pipets (5 per two students)
5 ml pipets (2 per two students)
test tubes (10 per two students)
test tube racks (1 per two students)
urine specimen jars (1 per two students)
100 or 250 ml graduated cylinders (1 per two students)
spectrophotometers and spectrophotometer tubes
clinical centrifuges and conical centrifuge tubes
70% alcohol
sterile cotton
sterile finger lancets
Pasteur pipets
0.01 ml Sahli blood pipets or 10 μl micropipets
solid anticoagulant crystals: heparin, sodium EDTA, or sodium citrate
distilled water
urea assay kit (1 per laboratory section; see Note)

Note: A very good Urea Assay Kit is supplied by the Sigma Chemical Co., P. O. Box 14508, St. Louis, MO 63178.

Procedures

Preparation of Standard Curve
1. Accurately pipet the following amounts of reagents into a series of test tubes.

Tube #	Standard Urea Solution[a]	Water	Urease in Buffer
1	0.10 ml	0	0.5 ml
2	0.08 ml	0.02 ml	0.5 ml
3	0.06 ml	0.04 ml	0.5 ml
4	0.04 ml	0.06 ml	0.5 ml
5	0.02 ml	0.08 ml	0.5 ml
6	0	0.10 ml	0.5 ml

[a] A standard urea solution (30 mg urea N/100 ml) is provided with the Sigma kit. The solution used here should have 6 mg urea N/100 ml, a 5-fold dilution of the Sigma standard. Note that 6 mg urea N/100 ml corresponds to 12.7 mg urea/100 ml (i.e., urea is 47% N).

Mix the tubes carefully after each addition. Add the urease last.

2. Let the reaction mixtures incubate for 15-20 minutes at room temperature or 5-10 minutes in a 37°C waterbath. During this incubation the urea will be completely hydrolyzed to carbon dioxide and ammonia.
3. After the incubation period, add the following reagents: 1.0 ml of phenol nitroprusside, 1.0 ml of hypochlorite, and 5.0 ml of water. **The nitroprusside and hypochlorite must be added with a rubber pipetting bulb. They should not be pipetted by mouth.**
4. Incubate the tubes for an additional 30 minutes at room temperature. The blue color that develops is a measure of the ammonia released in the reaction. Using a spectrophotometer, measure the absorbance of each sample at 570 nm versus a reagent blank (tube #6) and plot a standard curve. Is the absorbance directly proportional to the concentration of urea?

Assay of Urea in Urine and Blood Plasma
The preparation of a blood plasma sample for urea determination is the most difficult part of this exercise and will require extreme care on your part.

1. Disinfect your finger with 70% alcohol, lance it with a sterile lancet, and collect 6-8 drops of blood (a smaller volume will not suffice) in a conical centrifuge tube to which a tiny crystal of solid anticoagulant has been added. Centrifuge the tube for 5 minutes to separate the cells from the plasma. After the centrifugation, carefully remove as much plasma as possible with a Pasteur pipet and transfer it to another small tube. Use 0.010 ml (10 μl) samples of this plasma for the urea assay. It is strongly suggested that duplicate samples be assayed. A convenient method is to transfer the 10 μl sample with a Sahli blood pipet or a conventional 10 μl micropipet to an assay tube containing 0.090 ml of water. (Note that the blood plasma has been diluted 10-fold in this procedure.) This sample is now ready for assay.

2. Collect a urine sample. Record the volume of the sample and the time at which it was collected. Urine has a fairly high concentration of urea, so it must be diluted. Dilute the urine 1000-fold and remove 10 μl samples for assay.
3. Assay the plasma and urine samples as before, and calculate the urea concentrations per milliliter of fluid from your standard curve. Calculate the urea production in milliliters per minute from your data. Finally compute the renal clearance value for urea.

STUDY QUESTIONS

1. How can the renal clearance values for various substances be so different; for instance, urea versus inulin versus sodium? (*Hint*: Consider that a substance may be cleared from the blood in three different ways: (1) glomerular filtration, (2) glomerular filtration but with partial tubular reabsorption back into the blood, and (3) glomerular filtration and additional tubular excretion from the blood.) How would these processes affect the clearance values for different substances? Diodrast gives a clearance value of 750 ml/minute. How can you explain this? Why do you suppose that inulin is often the substance of choice in clinical testing procedures?

2. What is the glomerular filtration rate (GFR) and how could you determine it for the human kidney?

3. Another clinical test of renal function is the so-called retrograde pyelogram. How is this test performed and what does it indicate about the functioning of the kidneys?

4. Characterize each of the following disorders of the kidney and urinary system.
    a. Kidney stones (renal calculi).
    b. Glomerulonephritis.
    c. Pyelitis.
    d. Cystitis.
    e. Ptosis (floating kidney).
   Which of these disorders could be diagnosed by the renal function test you have studied in this exercise?

REFERENCES

Barger, A. C., and J. A. Herd. 1971. The renal circulation. *N. Eng. J. Med., 284*:482.
Crawhall, J. C., et al. 1967. The renal clearance of amino acids in cystinurea. *J. Clin. Invest., 46*:1162.
Donat, P. E., et al. 1970. Fourteen-hour concentrated urine osmolality as a renal failure test in children. *J. Urol., 104*:478.
Faulkner, W. R., and J. W. King. 1970. "Renal Function Tests." In N. W. Tietz (ed.), *Fundamentals of Clinical Chemistry*. Saunders, Philadelphia.
Jacobson, M. H., et al. 1962. Urine osmolality: a definitive test of renal function. *Arch. Intern. Med., 110*:85.
Lonsdale, K. 1968. Human stones. *Sci. Amer., 219*(6):104.

Pitts, R. F. 1968. *Physiology of the Kidney and Body Fluids,* 2nd ed. Year Book, Chicago.

Relman, A. S., and N. G. Levinsky. 1971. "Clinical Examination of Renal Function." In M. B. Strauss and L. G. Welt (eds.), *Diseases of the Kidney,* 2nd ed. Little, Brown, Boston.

Rouiller, C., and A. F. Muller (eds.). 1969. *The Kidney* (3 vols.). Academic Press, New York.

Sigma Technical Bulletin #640. 1974. *The Colorimetric Determination of Urea Nitrogen at 500–650 nm in Plasma, Serum or Urine.* Sigma Chemical Co., Saint Louis.

Smith, H. W. 1956. *Principles of Renal Physiology.* Oxford University Press, New York.

EXERCISE 23

Name _____

Laboratory Section _____ Date _____

RESULTS AND CONCLUSIONS

Standard Curve

Plot a curve of absorbance at 570 nm versus urea concentration.

Calculation of Renal Clearance for Urea

	Example	*Experimental Data*
Time bladder was emptied	2:00 P.M.	_____
Time urine sample was collected	3:00 P.M.	_____
Elapsed time	60 min	_____
Volume of urine in sample	120 ml	_____
Rate of urine production	2 ml/min	_____
Urea concentration in blood plasma	0.2 mg/ml	_____
Urea concentration in urine	900 mg/120 ml	_____
Urea excreted in urine per minute	15 mg/min	_____
Renal clearance value	75 ml cleared/min	_____

Sources of Error

Can you suggest several possible sources of error in this experiment? Consider the following: (1) Did you really empty your bladder in collecting the urine sample for analysis? How could you be certain to collect the entire amount of urine produced in a given time? (2) Does the renal excretion of urea depend upon the overall rate of urine production? If so, does this introduce an error? Suppose, for example, you had been concerned that you would not produce enough urine in the 1 hour interval so you drank a liter of water to increase the flow of urine. Would this affect the renal clearance of urea? (If so, any test of renal clearance would depend in part on what one had drunk before the test and would not represent a true measure of kidney activity.) The text by Smith [1956] will be helpful in answering this question.

288 *Exercise 23] Results and Conclusions*

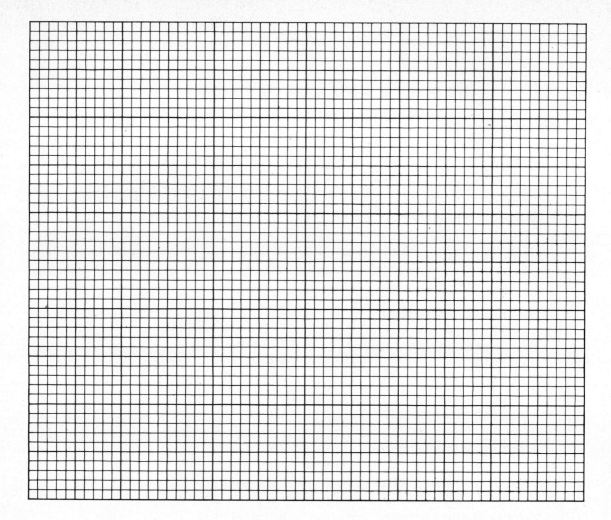

SECTION IX

Regulatory Processes

EXERCISE 24

Glucose Tolerance Test

OBJECTIVES In this exercise we shall perform a standard glucose tolerance test on ourselves. The initial blood glucose concentration and the values following ingestion of a standard amount of glucose will be related to data for normal and diabetic individuals.

The concentration of glucose within the blood of a normally healthy person falls within the limits of 65–120 mg/100 ml blood serum under ordinary circumstances. A blood sugar concentration greater than 120 mg % in the early morning at least 10 hours after a previous meal strongly indicates diabetes mellitus. In a standard glucose tolerance test, the subject drinks a solution containing a carefully weighed amount of sugar (50 grams). The rise in blood sugar levels after ingestion gives important information about the capacity of the body to regulate and maintain proper amounts of glucose in the blood, as well as the capacity of the blood to supply sugar to actively metabolizing tissues. Figure 24–1 presents typical data for two subjects in a standard test. The first subject has an initial blood sugar level of 80 mg % that rises rapidly to 140 mg % 1 hour after drinking the solution. Within 3 hours, however, the blood sugar falls to a subnormal hypoglycemic level. This reponse is typical of most people. The second subject has a high initial level that rises to 200 mg % after ingestion and even after 5 hours is abnormally high. This person is probably diabetic.

A convenient method for the quantitative assay of glucose in solution makes use of two enzymes and appropriate cofactors in a coupled sequence. Hexokinase and glucose-6-phosphate dehydrogenase catalyze the first two reactions outlined below.

$$\text{Glucose} + \text{ATP} \xrightarrow{\text{HK}} \text{Glucose-6-phosphate} + \text{ADP} \quad (1)$$

$$\text{Glucose-6-phosphate} + \text{NADP} \xrightarrow{\text{G-6-PD}} \text{6-Phosphogluconic acid} + \text{NADPH} \quad (2)$$

$$\text{NADPH} + \text{PMS} \longrightarrow \text{NADP} + \text{PMSH} \quad (3)$$

$$\text{PMSH} + \text{INT} \longrightarrow \text{PMS} + \text{INTH (red-colored formazan)} \quad (4)$$

Reactions (3) and (4) are nonenzymatic steps. NADPH generated in reaction (2) reacts with phenazine methosulfate (PMS) to produce PMSH. PMSH reacts with the oxidized tetrazolium dye (INT) to produce INTH, which is red in color. The amount of red color produced is directly proportional to the amount of glucose present in solution.

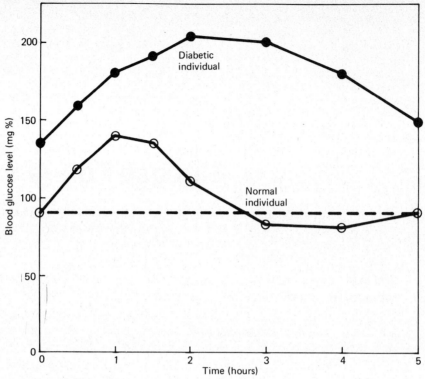

FIGURE 24-1 Time course of a standard glucose tolerance test in two subjects, one normal and one diabetic.

Materials

    drinking cups (1 per two students)
    glucose solution for drinking, exactly 50 grams of glucose dissolved in 200 ml water
    heparinized capillary tubes for collection of blood samples
    Seal-Ease
    hematocrit centrifuge
    12–15 ml test tubes (4 per group)
    test tubes (12 per group)
    test tube racks (1 per group)
    1 ml pipets (4 per group)
    5 ml pipets (1 per group)
    sterile lancets
    sterile cotton
    70% alcohol
    20 µl (0.2 ml) quantitative blood pipets, or 20 µl micropipets with suitable syringe control, or any pipets that can deliver 20 µl samples (1 per group)
    water baths (37°C)
    Gluco Strate Enzyme Reagent (15–20 ml per group; see Note)
    Gluco Strate Color Developer (3–4 ml per group; see Note)
    Calibrate 1, 2, and 3 (standard glucose solutions: 80, 150, and 300 mg/100 ml for calibration of the procedure; see Note)
    0.1 N HCl diluent (not provided with the assay kit) (50 ml per group)
    spectrophotometers and spectrophotometer tubes

Note: Available from General Diagnostics, Division of Warner-Lambert Company, Morris Plains, NJ 07950. A considerable advantage of the Gluco Strate reagent is that one need not deproteinize blood samples prior to assay.

Procedures

Glucose Assay Test

Calibration of the method may be performed at the time the blood sugar determinations are made, or it may be done beforehand to afford practice in the assay technique. Careful pipetting is essential for the success of this test.

1. Set up a series of test tubes. Include one for each glucose standard and a reagent blank. Greater accuracy may be obtained if each standard is run in duplicate.
2. To each tube add 1.0 ml of reconstituted Enzyme Reagent. Then add 20 μl of each standard solution to the appropriate tube, and mix.
3. Place all tubes, including the reagent blank, in a 37°C waterbath and allow the tubes to preincubate for about 3 minutes.
4. Pipet 0.2 ml of Color Developer into each tube and mix each thoroughly after the addition. Incubate all tubes at 37°C for at least 10 minutes. (Precise timing for the addition of reagents is not necessary if the assay is run for 10 minutes or more. Prolonged incubation beyond 10 minutes produces a higher reagent blank.)
5. Add 5.0 ml of 0.1 N HCl (diluent) to each tube after 10 minutes, and mix. This addition stops the reaction.
6. Within 45 minutes, use a suitable spectrophotometer to determine the absorbance of each sample at wavelength 500 nm. Use the reagent blank to zero the spectrophotometer.

Blood Sugar Determination

Determination of the blood sugar level requires careful and accurate pipetting. If laboratory time does not permit the class to perform a complete glucose tolerance test, it is entirely possible to perform a single measurement of blood glucose. This value in itself has diagnostic significance.

1. The test subject cleanses one of his fingers with 70% alcohol and punctures it with a sterile lancet. Blood is collected in a heparinized capillary tube (as described in Exercise 1 on blood chemistry).
2. The capillary tubes are closed off at one end with Seal-Ease and spun for 5 minutes in a hematocrit centrifuge.
3. After the centrifugation, the clear serum (supernatant) is collected from the capillary tube and 20 μl samples are used for blood sugar determination. The capillary tube can be broken by hand and the fluid collected in a clean tube. Prior scratching of the capillary tube with a small file may assist in breaking the tube.

Glucose Tolerance Test

1. A convenient procedure is to divide students into groups of four. One individual with no history of diabetes or with a known prediabetic condition serves as the subject. Another student serves as sample taker and overseer. The other two students are the glucose analyzers. All four should thoroughly familiarize themselves with the details of each step of the procedure.

2. The subject should not have eaten within a 3 hour period prior to the test. Preferably, he should have skipped the previous meal altogether. Have the subject drink a sufficient volume of glucose solution to provide 1 gram of glucose per kilogram of body weight. Take a blood sample immediately and assay it for blood sugar. At ½ hour intervals, remove samples over a period of at least 1½ hours from time 0, and assay each for blood sugar. If time permits, a 2 hour sample will be helpful.

STUDY QUESTIONS

1. Before the discovery of insulin, it was suspected that the pancreas might be responsible for the synthesis of an antidiabetic factor. An early experimental approach to this problem was to prepare an extract of whole animal pancreas and inject this into diabetic dogs. Unfortunately this method did not yeild positive results. Can you suggest a reason for this failure? (*Hint:* Consider the chief functions of the pancreas.) What modifications of this procedure did the Canadian researchers Banting and Best employ in their discovery of insulin?

2. In what form is sugar stored in the body and what are the chief regions of storage? How is insulin thought to regulate sugar metabolism in the body? How is the production of insulin in the pancreas regulated? What is the physiological basis of the hypoglycemic response depicted in Figure 24-1? (Several of the references at the end of this exercise will be helpful in dealing with these questions.)

3. Can one take insulin orally? What are the oral medications and how are they thought to act? What are modified insulins and why were they developed?

4. What is the primary structure of insulin? What is proinsulin? How is the conversion of proinsulin to insulin brought about? What is the physiological significance of this transformation?

5. What is the commercial source of insulin for diabetics? Why isn't this material antigenic?

6. Insulin is a protein composed of 51 amino acids and consists of two polypeptide bonds crosslinked by disulfide bridges. Suppose that you wanted to synthesize insulin chemically by linking together free amino acids, adding one at a time. Suppose further that you got a 75% yield for each step of the synthesis. What percent yield would you have for the overall synthesis (given the presence of 49 peptide bonds in one molecule of insulin)? Does this mean that chemical synthesis is not feasible as a potential source of insulin?

REFERENCES

Banting, F. G., and C. H. Best. 1922. The internal secretion of the pancreas. *J. Lab. Clin. Med.,* 7:251.

Banting, F. G., and C. H. Best. 1922. Pancreatic extracts. *J. Lab. Clin. Med.,* 7:464.

Curry, D. L., et al. 1968. Dynamics of insulin secretion by the perfused rat pancreas. *Endocrinology,* 82:572.

Davidoff, F. F. 1968. Oral hypoglycemic agents and the mechanism of *diabetes mellitus. New Eng. J. Med., 278*:148.

Fajans, S. S., and J. W. Conn. 1959. The early recognition of *diabetes mellitus. Ann. N. Y. Acad. Sci., 82*:208.

Fajans, S. S. 1971. What is diabetes? *Med. Clin. North Am., 55*:821.

Ganong, W. F. 1973. *Review of Medial Physiology,* 6th ed. Lange Medical Publications, Los Altos, CA. (Contains a discussion of the hypoglycemic action of modified insulins, including time of peak action and duration of action, p. 252. Also discusses oral medications, p. 260.)

Goeschke, H., et al. 1974. Circadian variation of carbohydrate tolerance in mature onset diabetics treated with sulfonylureas. *Horm. Metab. Res., 6*:386.

Klimt, C. R., et al. 1968. Standardization of oral glucose tolerance test: Report of the committee on statistics of the American Diabetes Association. *Diabetes, 18*:299.

Levine, R. 1970. Mechanics of insulin secretion. *New Eng. J. Med., 283*:522.

Merrifield, R. B. 1969. Solid-phase peptide synthesis. *Adv. Enzymol., 32*:221.

EXERCISE 24

Name _____

Laboratory Section _____ Date _____

RESULTS AND CONCLUSIONS

Standard Glucose Curve

Plot a standard curve relating absorbance at 500 nm to the glucose concentration. Why is this wavelength used? Is the absorbance proportional to the glucose concentration?

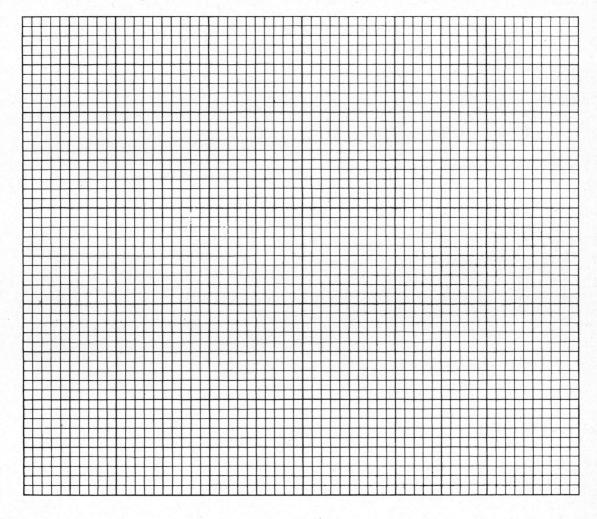

Exercise 24] Results and Conclusions

Blood Sugar Concentration

Time of day at which assay is performed_____

Number of elapsed hours since your last meal_____hours

Blood sugar concentration_____mg %

Glucose Tolerance Test

Plot a graph of blood sugar level against time in hours after drinking the sugar solution. How much did your blood sugar increase one hour after drinking the solution? What can you conclude from your curve? Do you show a hypoglycemic response?

Exercise 24] Results and Conclusions

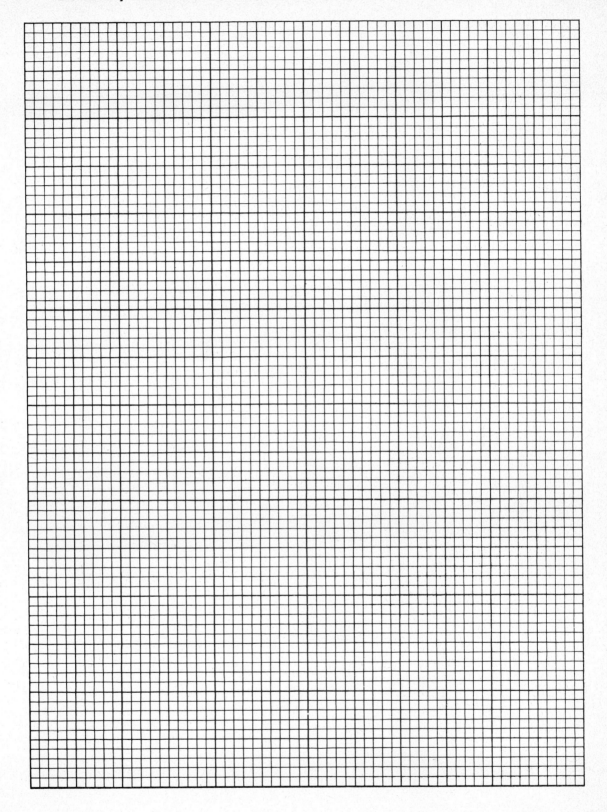

EXERCISE 25

Regulation of Body Temperature: The Use of Radiotelemetry

OBJECTIVES In this exercise we shall investigate several basic aspects of temperature regulation in man, including (1) the microscopic anatomy of the skin; (2) the location of sweat glands on several regions of the body; and (3) our own body temperature and its regulation using simple radiotelemetric equipment. Experiments to investigate the control of body temperature during exercise and the changes of body temperature in poikilothermic organisms are also suggested.

Animals are often characterized as being cold-blooded or warm-blooded. Cold-blooded animals, which include all invertebrate organisms, fishes, amphibians, and reptiles, generate body heat, but are unable to vary their production of warmth in response to cold or to maintain a constant internal temperature by conserving bodily warmth. Although many of these animals achieve some measure of control behaviorally (by seeking warmer or cooler regions in their habitat), they are best described as being poikilothermic ("variable in temperature") in that their internal temperature is largely dependent upon that of the surroundings. These organisms are also termed ectothermic.

Man and the mammals, as well as all birds, maintain a nearly constant internal temperature and are said to be homeothermic ("same temperature") or endothermic. Warming processes depend on an increase in the rate of metabolism; cooling processes usually depend on the evaporation of water from the surface of the body. Temperature control consists, in effect, of a balance between warming and cooling processes. The hypothalamus serves as the chief regulatory center. Many mammals sweat while others pant to facilitate evaporation from the wet surfaces of respiratory membranes. Several small rodents spread saliva over their skin and body fur to effect surface evaporation. In addition, internal physiological processes such as changes in the rate of blood flow within the body assist in temperature regulation.

The water of the blood and tissue fluids possesses three attributes which make it a preeminently useful medium for temperature control. The specific heat of water is considerably higher than that of any other fluid. An average man produces heat equivalent to 3000 kcal/day, but this is capable of raising the body temperature only about 30°C because of the high specific heat of water. In his classic treatise, *The Fitness of the Environment,* L. J. Henderson pointed out that if the body tissues had the low heat storage capacity of most substances, this amount of heat would raise the temperature of the tissues and body fluids by 100–150°C.

The high thermal conductivity of water assists in the transfer of warmth in the body and the high latent heat of evaporation means that the evaporation of a relatively small volume of moisture results in great heat loss. These properties of water are obviously of significance in the dissipation of excess body warmth.

MICROSCOPIC ANATOMY OF THE SKIN

Materials

microscope slides of human skin showing sweat glands and a variety of nerve endings

Procedures

1. Carefully draw and label one or more sweat glands. In which skin layer(s) are the glands located?
2. Can you observe warmth or cold receptors in the skin? If so, include these in your drawing.

DISTRIBUTION OF SWEAT GLANDS

There are over 2½ million sweat glands distributed over the surface of the human body. These are of prime importance in regulating body temperature. Anhidrosis, an inherited condition (sex-linked recessive) that results in the total absence of sweat glands in men who inherit a single abnormal allele, is an extremely serious and often fatal condition.

The position of the sweat glands on various parts of the human body can be mapped very accurately. The procedure involves painting iodine onto the skin, allowing it to dry and pressing bond paper onto the painted square. Each tiny drop of sweat forming in the skin absorbs the iodine and reacts with starch in the paper to produce a purple spot.

Materials

1% iodine solution
bond paper
scissors

Procedures

1. Paint a 1 inch square with 1% iodine on the following parts of your body: palm of the hand, forehead, sole of the foot, back of the neck, and the subcubital region. Allow the iodine to dry thoroughly.
2. Cut out 1 inch squares of bond paper. Press one square to each iodine spot for about 30 seconds. Remove the paper and count the number of purple dots in 1

cm^2. This number represents the number of sweat glands actively secreting at the time of your measurement.
3. Repeat these procedures after you have exercised sufficiently to produce a mild sweat. You may also repeat the determinations after exposure to cold for a short period of time in a cold room or outdoors. Tabulate and interpret your results.

RADIOTELEMETRIC DETERMINATION OF BODY TEMPERATURE

Detailed methodology for the study of temperature regulation in both poikilotherms and homeotherms has been devised [Stockdale, 1975]. This procedure uses an extremely small temperature-sensitive radio transmitter to monitor body temperature. The radio transmitter contains a temperature-sensitive element (a thermister) that transmits "clicks" at a rate proportional to its temperature. A standard AM radio receiver is used to detect the low strength radio signals, or slightly more expensive transmitters tuned to particular frequencies are monitored in the shortwave range. The telemeters are calibrated by immersing them in warm water and counting the clicks as the water cools. Suitable calibration curves are constructed.

By means of radiotelemetry one can measure the surface temperature of the human body quickly and accurately under a variety of conditions.

Materials

mercury thermometers
test tubes
250 ml beakers
Bunsen burners and ringstands
Mini-Mitter radio transmitter(s) (see Note)
AM radio receiver(s)

Note: Temperature-sensitive radio transmitters can be purchased very inexpensively from the Mini-Mitter Co., Inc., P. O. Box 88210-G, Indianapolis, IN 46208.

Procedures

Calibration of the Mini-Mitter
1. Calibration is done by preparing a beaker of 40°C water and placing the Mini-Mitter inside the bath. As the temperature cools the number of clicks per 30 seconds as received in the AM range is counted and recorded for a series of temperatures. A calibration curve from 36° to 40°C is prepared. (A broader range can be established if the procedures with poikilothermic organisms described later in this exercise are contemplated.) As the calibration process is somewhat time consuming, it may well be done prior to the regular laboratory period.
2. Place the transmitter on your lab bench and determine the number of clicks for three 30 second intervals. Average them and verify the ambient temperature with a thermometer.

Body Temperature in Man

1. Place the transmitter inside a suitable plastic wrapper and position the device under your tongue. The plastic wrapper should be long enough to come out of your mouth, so that you don't swallow the whole works! Record your oral temperature.
2. Determine the temperature under your armpit. You should perform this measurement with your arm both in an abducted position and securely adducted.
3. Measure the surface temperature of your skin in the region of the groin. Stockdale suggests that this be done in a restroom and that the transmitter be placed against the skin of the groin about 1 inch down the thigh. Try it with your legs crossed and uncrossed.
4. Choose other regions of the body for further measurements. The Mini-Mitter can be secured in place with adhesive tape.

Body Temperature During Exercise

Design and carry out an experiment to study body temperature during exercise. The nature of this will depend largely on the available time and facilities. Craig and Abraham [1974] suggest that body temperature and heart rate be monitored periodically with a rectal thermometer during a basketball game; similar investigations with the bicycle ergometer or other exercise equipment are possible. A simple suggestion is to study body temperature in an athletically inclined student who runs laps in the gymnasium. Plan and execute your investigation carefully and make other physiological measurements that you feel may be important. Oral and rectal thermometers, or the Mini-Mitter in a plastic wrapper under the tongue may be used.

Changes of Body Temperature in Poikilotherms

A study of temperature changes in poikilothermic organisms offers an interesting and valuable contrast to the study of homeothermy. Several investigators have outlined laboratory procedures for the study of temperature regulation in amphibians and reptiles using the Mini-Mitter. In these investigations the reptile or amphibian swallows the Mini-Miter and the body temperature is measured as the ambient temperature is altered, or the animal is placed in a temperature gradient and its temperature preference is noted. The procedures devised by Osgood [1971] and Berry [1971] offer several possibilities for laboratory study, and can provide the basis for independent study work in physiology. The experiments require the participation of someone with practical experience in the handling of amphibians and reptiles and should not be undertaken without a skilled handler. An important consideration is the retrieval of the radiotelemeter without causing harm to the animal. A skilled handler can retrieve the instrument by applying gentle pressure to the animal's abdomen to induce regurgitation. Smaller amphibians and reptiles are not suitable for this investigation.

STUDY QUESTIONS

1. Outline in detail the various mechanisms of heat loss and heat production in the human body. What are so-called active and passive mechanisms of heat loss?

2. How does the hypothalamus regulate the conservation or loss of heat in the body? What is the experimental evidence underlying your contentions?

REFERENCES

Benziger, T. H. 1961. The human thermostat. *Sci. Amer., 204*(1):134.

Berry, J. W. 1971. The study of animal temperature regulation by telemetry. (Available without charge from the Mini-Mitter Company.)

Bogert, C. M. 1959. How reptiles regulate their body temperature. *Sci. Amer., 200*(4):105.

Craig, A. B., and W. M. Abraham. 1974. Student laboratory exercise in temperature regulation. *Physiol. Teacher, 3*(2):1.

Dawkins, M. J. R., and D. Hull. 1965. The production of heat by fat. *Sci. Amer., 213*(2):62.

Hardy, J. D. 1969. Physiology of temperature regulation. *Physiol. Rev., 41*:521.

Irving, L. 1966. Adaptations to cold. *Sci. Amer., 214*(1):94.

Mackay, R. S. 1968. *Bio-Medical Telemetry: Sensing and Transmitting Information from Animals to Man.* Wiley, New York.

Montagna, W. 1965. The skin. *Sci. Amer., 212*(2):56.

Osgood, D. W. 1971. Reptilian behavioral thermoregulation studied by telemetry. (Available without charge from the Mini-Mitter Company.)

Stockdale, D. L. 1975. Radio-telemetry temperature studies of homoiotherms and poikilotherms. *Physiol. Teacher, 4*(3):1.

Name _____

Laboratory Section _____ Date _____

EXERCISE 25

RESULTS AND CONCLUSIONS

Anatomy of the Skin

Make careful drawings of one or more sweat glands. In which skin layer(s) are the glands located?

Distribution of Sweat Glands

Summarize your findings on the number and location of sweat glands on various parts of the human body. What are the effects of exposure to warmth and cold on the amount of perspiration and on the number of glands actively secreting?

308 *Exercise 25] Results and Conclusions*

Radiotelemetry

Draw a standard curve relating temperature and the number of clicks emitted by the Mini-Mitter.

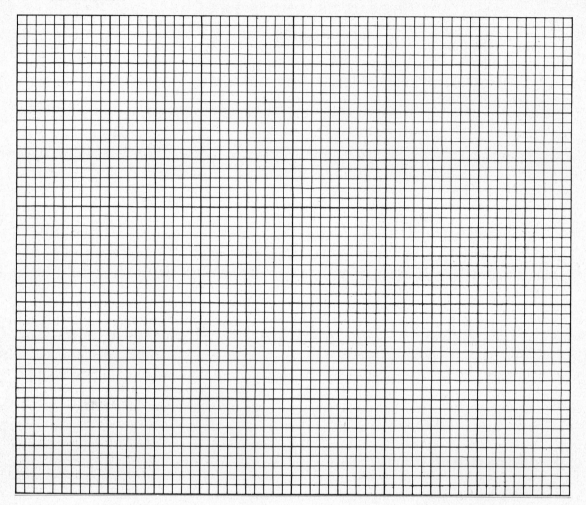

Summarize your findings on body temperature in man, the effects of exercise on body temperature, and any other simple experiments you undertake. Interpret your findings.

EXERCISE 27

Circadian Rhythms in Man

OBJECTIVES In this exercise we shall study variations in heart and respiratory rate, body temperature, blood pressure, eye-hand coordination, and hand grip strength at specific intervals over a period of several days. Results of the Horne-Östberg questionnaire concerning daily activity rhythms are to be correlated with the physiological data. The findings of this exercise are to be evaluated as possible circadian rhythms in man.

Rhythmical processes play a highly significant role in the normal life activities of all organisms. The physiology and overt behavioral responses of plants, animals, and man are greatly influenced by a rhythm with a periodicity approximately but not exactly equal to 24 hours; hence the name *circadian (circa,* approximately; *dies,* day) rhythms. Several early studies demonstrated that bean plants exhibit periodic leaf movements (raising and lowering the leaves once a day) that persist even under conditions of constant temperature and illumination. This endogenous leaf movement was later shown to have a period length of *about* 24 hours under free-running conditions (that is, in the absence of external synchronizers), but could be entrained to an *exact* 24 hour periodicity with appropriate light/dark signals. Similar rhythms have been found in unicellular flagellates, many plant species, fruit flies, crabs, and a wide variety of invertebrate organisms, all classes of vertebrates, and man.

The chief characteristics of circadian rhythms may be summarized as follows:

(1) The free-running period under constant environmental conditions approximates but is ordinarily not equal to 24 hours.
(2) The period length may be entrained by external cues (synchronizers) to a period length of exactly 24 hours.
(3) The rhythmical behavior may be phase shifted (either advanced or delayed) through experimental manipulation.
(4) The period length is temperature insensitive in that an increase of 10°C does not increase the frequency of oscillation by a factor of 2 or 3 (as for ordinary chemical processes); indeed, the Q_{10} in many cases approximates one.

Circadian rhythms have been discovered in man that affect a wide variety of physiological processes. Most people know that body temperature reaches a high point in the late afternoon and a low in the early morning. There is also clear evidence for maximum and minimum points within the daily cycle for body temperature, water excretion, potassium and calcium ion elimination, the presence of steroids in the urine, the number of actively dividing cells in a particular tissue, reaction to a drug, and the speed and accuracy with which mental computations can be carried out. Indeed, there is no single organ in the body that does not show rhythmical activity. Cellular and molecular phenomena such as enzyme activity, osmotic pressure, respiration rate, and membrane permeability are also subject to periodic fluctuations in a 24 hour cycle.

Without further evidence, we could not be certain that these diurnal variations fit the criteria for true circadian behavior. However, research performed on human volunteers placed in rigorously maintained constant conditions has shown that many daily fluctuations are truly circadian. Aschoff's studies on student volunteers kept in isolation, for example, demonstrated that urine excretion, body temperature, and overall activity assumed a periodicity greater than 24 hours in a large proportion of the cases studied (22 of 26 subjects showed a "day" length of about 25 hours for each of these parameters), but that the periodicities could be entrained to a normal cycle by reintroducing ordinary 24 hour light/dark signals.

Physicians are becoming increasingly aware that the proper dosage of a drug may be quite different at different times of day; in some cases, what constitutes a beneficial dose of a pharmacological agent may actually be deleterious when given at an inappropriate time in the cycle. Experiments with bacterial toxins have shown that these substances may kill a vast majority of experimental animals at one time of day, but have little if any effect when administered at another hour.

A further problem exists with respect to rapid transcontinental flight. A person may be physically present in a new locality, but his rhythms behave for several days as though he were still at the point of origin. In addition, it has been suggested that when a person is exposed to conditions that shift the phase of his "biological clocks," the individual rhythms of the various organs do not necessarily shift together; some may adjust fairly rapidly, while others may require a week or so to readjust. Such uncoordinated shifts may well lead to serious psychological disturbances and even underlie pathological states in the body. Obviously, circadian rhythms form an extremely important field of biomedical research, and one that will become increasingly significant.

One of the very interesting aspects of research on rhythms centers on individual differences in various people. Popular wisdom acknowledges the existence of "morning" and "evening" types, and scientific study has confirmed and extended this view. Many investigators have correlated daily fluctuation of certain physiological parameters with work efficiency, manual dexterity, intellectual clarity, and analytical ability, as well as with subjective assessment of "how one feels."

In this exercise we shall measure certain diurnal fluctuations in our own physiological activities and examine the evidence that these variations constitute true circadian rhythms. In addition, each student will answer a "morningness/eveningness" questionnaire dealing with his activity rhythm and attempt to correlate the conclusions of this questionnaire with the physiological data derived from the measurements. Students interested in pursuing this topic will find two popular books, *Body Time* [Luce, 1971] and *The Living Clocks* [Ward, 1971], extremely useful and interesting introductions. "Circadian Rhythms in Man" [Aschoff, 1965] cites very interesting findings on human subjects confined under constant-environment conditions.

Exercise 27]   Circadian Rhythms in Man

Materials

oral thermometer
sphygmomanometer and stethoscope
hand dynamometer
stopwatch or clock with secondhand sweep
wooden ruler (or Nelson Reaction Timer)

Procedures

This exercise is a take-home laboratory. Materials are to be checked out for the necessary length of time, and measurements made every 6 hours (preferably every 4 hours) for a minimum of three days. If possible, readings during the period 4:00–8:00 P.M. should be made more frequently. Readings in the night are made at 3:00 or 4:00 A.M. The subject should set the alarm and allow himself 10–15 minutes to awaken. The experimental determinations obviously require the assistance of a roommate or spouse.

1. *Sitting heart rate.* After 5 minutes of quiet sitting, measure the heart rate by palpation of the radial artery. Record the results.
2. *Sitting respiratory rate.* Count the number of respirations per minute at the same time the heart rate is taken. Record the results.
3. *Sitting blood pressure.* Immediately following the determination of heart rate, measure blood pressure and record the reading as systolic/diastolic.
4. *Oral temperature.* Record the oral temperature to the nearest tenth of a degree. Be certain that the thermometer has been present in the mouth long enough to equilibrate.
5. *Eye-hand coordination.* Carry out the eye-hand coordination test as described below. The roommate (spouse) should administer the test 10 times. The subject seats himself with his arm resting comfortably on a table top and his preferred hand extending just over the table's edge. The subject opens his thumb and index finger to a distance of one inch. The experimenter positions the ruler directly over the two fingers (with the bottom of the ruler at finger level) and releases the ruler a few seconds after announcing the command "Ready!" The subject catches the ruler between his two fingers as quickly as he can, and the distance is recorded. The data are recorded as the mean reaction distance.
6. *Hand grip strength.* The subject is to squeeze the hand dynamometer maximally with the dominant hand three times. The subject should not be permitted to see the scores of this test as this may influence the results. Use the highest score as the value.

Answer the Horne-Östberg "morningness/eveningness" questionnaire on the following pages and score the results according to the key provided in Appendix B.

HORNE-ÖSTBERG
MORNINGNESS/EVENINGNESS QUESTIONNAIRE

Instructions

(1) Please read each question very carefully before answering.
(2) Answer ALL questions.
(3) Answer the questions in numerical order.
(4) Each question should be answered independently of others. Do NOT go back and check your answers.
(5) All questions have a selection of answers. For each question place an X alongside ONE answer only. Some questions have a scale instead of a selection of answers. Place an X at the appropriate point along the scale.
(6) Please answer each question as honestly as possible.

1. Considering only your own "feeling best" rhythm, at what time would you get up if you were entirely free to plan your day?

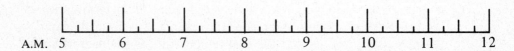

2. Considering only your own "feeling best" rhythm, at what time would you go to bed if you were entirely free to plan your evening?

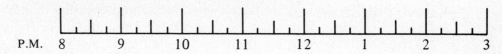

3. Is there a specific time at which you have to get up in the morning? To what extent are you dependent on being awakened by an alarm clock?

   not at all dependent   _____

   slightly dependent   _____

   fairly dependent   _____

   very dependent   _____

4. Assuming adequate environmental conditions, how easy do you find getting up in the morning?

   not at all easy   _____

   not very easy   _____

fairly easy _____

very easy _____

5. How alert do you feel during the first half-hour after waking up in the morning?

   not at all alert _____

   slightly alert _____

   fairly alert _____

   very alert _____

6. How is your appetite during the first half-hour after waking up in the morning?

   very poor _____

   fairly poor _____

   fairly good _____

   very good _____

7. During the first half-hour after waking in the morning, how tired do you feel?

   very tired _____

   fairly tired _____

   fairly refreshed _____

   very refreshed _____

8. When you have no commitments the next day, at what time do you go to bed compared to your usual bedtime?

   seldom or never later _____

   less than 1 hour later _____

   1–2 hours later _____

   more than 2 hours later _____

9. You have decided to engage in some physical exercise. A friend suggests that you do this 1 hour twice a week, and the best time for him is between 7 and 8 A.M. Bearing in mind nothing else but your "feeling best" rhythm, how do you think you would perform?

   would be in good form _____

   would be in reasonable form _____

   would find it difficult _____

   would find it very difficult _____

10. At what time in the evening do you feel tired and, as a result, in need of sleep?

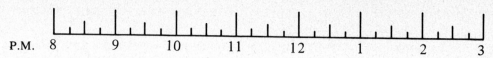

11. You wish to be at your peak performance for a 2 hour test that you know is going to be mentally exhausting. You are entirely free to plan your day. Considering only your own "feeling best" rhythm, which ONE of the four testing times would you choose?

    8:00–10:00 A.M. _____

    11:00–1:00 P.M. _____

    3:00–5:00 P.M. _____

    7:00–9:00 P.M. _____

12. If you went to bed at 11:00 P.M., at what level of tiredness would you be?

    not at all tired _____

    a little tired _____

    fairly tired _____

    very tired _____

13. For some reason you have gone to bed several hours later than usual, but there is no need to get up at any particular time the next morning. Which ONE of the following events are you most likely to experience.

    will wake up at usual time and will NOT fall asleep again _____

    will wake up at usual time and will doze thereafter _____

    will wake up at usual time but will fall asleep again _____

    will not wake up until later than usual _____

14. One night you have to remain awake between 4 and 6 A.M. in order to carry out a night watch. You have no commitments the next day. Which ONE of the following alternatives will suit you best?

    would NOT go to bed until watch was over _____

    would take a nap before and sleep after _____

    would take a good sleep before and nap after _____

    would take ALL sleep before watch _____

15. You have to do 2 hours of hard physical work. You are entirely free to plan your day. Considering only your own "feeling best" rhythm, which ONE of the following times would you choose?

Exercise 27]   Circadian Rhythms in Man

8:00–10:00 A.M. _____

11:00–1:00 A.M. _____

3:00–5:00 A.M. _____

7:00–9:00 P.M. _____

16. You have decided to engage in hard physical exercise. A friend suggests that you do this for 1 hour twice a week, and the best time for him is between 10 and 11 P.M. Bearing in mind nothing else but your own "feeling best" rhythm, how well do you think you would perform?

would be in good form _____

would be in reasonable form _____

would find it difficult _____

would find it very difficult _____

17. Suppose that you can choose your own work hours. Assume that you worked a 5 hour day (including breaks), and that your job was interesting and paid by results. Which FIVE CONSECUTIVE HOURS would you select?

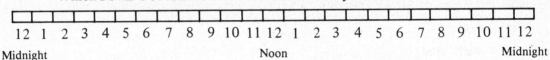

12 1 2 3 4 5 6 7 8 9 10 11 12 1 2 3 4 5 6 7 8 9 10 11 12
Midnight   Noon   Midnight

18. At what time of the day do you think that you reach your "feeling best" peak?

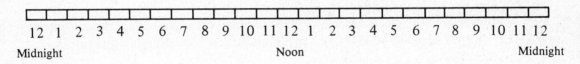

12 1 2 3 4 5 6 7 8 9 10 11 12 1 2 3 4 5 6 7 8 9 10 11 12
Midnight   Noon   Midnight

19. One hears about "morning" and "evening" types of people. Which ONE of these types do you consider yourself to be?

definitely a "morning" type _____

rather more a "morning" than an "evening" type _____

rather more an "evening" than a "morning" type _____

definitely an "evening" type _____

STUDY QUESTIONS

1. Rapid transcontinental flight across several time zones brings about "desynchronization" of one's circadian rhythms from one environment to another. Would you expect east to west flight to be more disruptive than west to east? Why? How long does readjustment take? Discuss Aschoff's findings on phase shifting in the European chaffinch and the possible significance these results have for the human condition.

2. In one of Aschoff's isolation experiments with human volunteers [1965], a particular subject showed a free-running rhythm of body temperature and several other physiological parameters with a period length of 24.7 hours. Activity (sleeping and waking) showed a very different rhythm, however, with a "day" length of 32.6 hours. What do you expect the consequences of this desynchronization and dissociation might be over a period of time? In all of Aschoff's studies the subjects took careful notes daily on how they felt.

3. What is the role of the pineal gland in controlling circadian rhythms?

4. One of the early studies on rhythms by the Swedish investigator Forsgren [1928] showed that liver activities exhibit very definite daily fluctuations. Hepatic function oscillates between a *secretory* phase (with a maximum around 3 P.M.) in which bile is produced and an *assimilatory* phase (with a maximum around 3 A.M.) in which large amounts of glycogen are stored. What is the significance of this finding with regard to normal human physiology? How would this rhythm be affected in a night worker who slept in the day? What possible significance do Forsgren's findings have for medical diagnosis and therapy?

5. What is Aschoff's rule regarding free-running period and incident light intensity? Is the human being "day active" or "night active" according to this rule?

REFERENCES

Aschoff, J. 1965. Circadian rhythms in man. *Science, 148*:1427.

Aschoff, J. (ed.). 1965. *Circadian Clocks.* North-Holland, Amsterdam.

Aschoff, J. 1969. Desynchronization and resynchronization of human rhythms. *Aerospace Med., XL*:844.

Brown, F. A., Jr., J. W. Hastings, and J. D. Palmer. 1970. *The Biological Clock, Two Views.* Academic Press, New York.

Bünning, E. 1973. *The Physiological Clock,* 3rd ed. Springer-Verlag, New York.

Colquhoun, W. P. (ed.). 1971. *Biological Rhythms and Human Performance.* Academic Press, New York.

Forsgren, E. A. 1928. On the relationship between the formation of bile and glycogen in the liver of rabbits. *Skandin. Arch f. Physiol., LIII*:137.

Halberg, F., et al. 1972. Autorhythmometry procedures for physiologic self-measurements and their analysis. *Physiol. Teacher, 1*(4):1.

Horne, J. A., and O. Östberg. 1976. A self-assessment questionnaire to determine morningness-eveningness in human circadian rhythms. *Int. J. Chronobiology,* 4:97.

Luce, G. G. 1971. *Body Time: Physiological Rhythms and Social Stress.* Bantam Books, New York.

Richter, C. P. 1965. *Biological Clocks in Medicine and Psychiatry.* Charles C Thomas, Springfield, IL.

Sollberger, A. 1965. *Biological Rhythm Research.* Elsevier, Amsterdam.

Ward, R. R. 1971. *The Living Clocks.* Knopf, New York.

Wright, E. 1959. Factors influencing diurnal variation of grip strength. *Res. Quart., 30*:110.

EXERCISE 27

Name _____

Laboratory Section _____ Date _____

RESULTS AND CONCLUSIONS

1. Record the data for the various parameters you have studied in a table such as the following:

Parameter Tested	Results 3/28/76 8 A.M.	Results 3/28/76 12 A.M.
Heart rate (seated)	78/minute	82/minute
Respiratory rate (seated)	14/minute	15/minute
Blood pressure (seated)	120/80	122/82
Oral temperature	37.5°C	37.8°C
Eye-hand coordination	13.5 cm	14.0 cm
Hand grip strength	60 pounds	65 pounds

2. Plot graphs to illustrate the fluctuation of these parameters over the course of several days. Are there maxima and minima? When do they occur?

3. Are the fluctuations you have observed truly circadian, or are they simple responses to periodic variations in the environment? What arguments can you advance in support of your contentions? Several references at the end of this exercise will be helpful in evaluating this question.

4. What is your score on the Horne-Östberg questionnaire? What type are you? Collect the class data and examine them to see if correlations can be drawn between the physiological data measured in the experimental portion of the exercise and the results of the questionnaire. What conclusions can you draw? (You may wish to compare the physiological data between extreme "morning" and "evening" types in the class.)

Exercise 27] Results and Conclusions

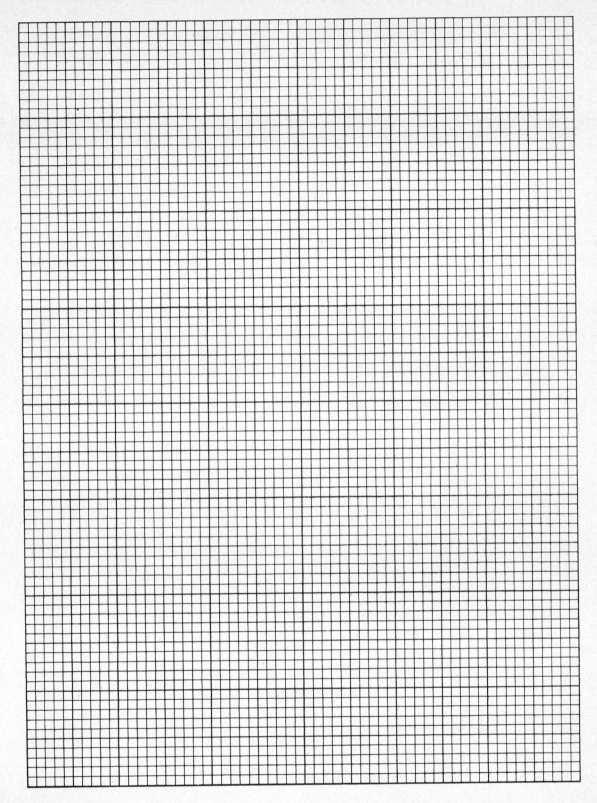

Appendix

A. BODY SURFACE AREA IN METERS SQUARED[a]

Height (cm)	\multicolumn{17}{c}{Weight (kg)}
	25	30	35	40	45	50	55	60	65	70	75	80	85	90	95	100	105
200							1.84	1.91	1.97	2.03	2.09	2.15	2.21	2.26	2.31	2.36	2.41
195						1.73	1.80	1.87	1.93	1.99	2.05	2.11	2.17	2.22	2.27	2.32	2.37
190				1.56	1.63	1.70	1.77	1.84	1.90	1.96	2.02	2.08	2.13	2.18	2.22	2.28	2.33
185				1.53	1.60	1.67	1.74	1.80	1.86	1.92	1.98	2.04	2.09	2.14	2.19	2.24	2.29
180				1.49	1.57	1.64	1.71	1.77	1.83	1.89	1.95	2.00	2.05	2.10	2.15	2.20	2.25
175	1.19	1.28	1.36	1.46	1.53	1.61	1.67	1.73	1.79	1.85	1.91	1.96	2.01	2.06	2.11	2.16	2.21
170	1.17	1.26	1.34	1.43	1.50	1.57	1.63	1.69	1.75	1.81	1.86	1.91	1.96	2.01	2.06	2.11	
165	1.14	1.23	1.31	1.40	1.47	1.54	1.60	1.66	1.72	1.78	1.83	1.88	1.93	1.98	2.03	2.07	
160	1.12	1.21	1.29	1.37	1.44	1.50	1.56	1.62	1.68	1.73	1.78	1.83	1.88	1.93	1.98		
155	1.09	1.18	1.26	1.33	1.40	1.46	1.52	1.58	1.64	1.69	1.74	1.79	1.84	1.89			
150	1.06	1.15	1.23	1.30	1.36	1.42	1.48	1.54	1.60	1.65	1.70	1.75	1.80				
145	1.03	1.12	1.20	1.27	1.33	1.39	1.45	1.51	1.56	1.61	1.66	1.71					
140	1.00	1.09	1.17	1.24	1.30	1.36	1.42	1.47	1.52	1.57							
135	0.97	1.06	1.14	1.20	1.26	1.32	1.38	1.43	1.48								
130	0.95	1.04	1.11	1.17	1.23	1.29	1.35	1.40									
125	0.93	1.01	1.08	1.14	1.20	1.26	1.31	1.36									
120	0.91	0.98	1.04	1.10	1.16	1.22	1.27										

[a] Reprinted from E. F. DuBois, *Basal Metabolism in Health and Disease,* 3rd ed. Philadelphia, Lea & Febiger, 1936. Used by permission.

B. KEY FOR EVALUATION OF THE HORNE-ÖSTBERG MORNINGNESS/EVENINGNESS QUESTIONNAIRE

Scoring

For questions 3-9, 11-16, and 19, the appropriate score for each response is displayed beside the answer box. For questions 1, 2, 10, and 18, the X made along each scale is referred to the appropriate score value range below the scale. For question 17 the most extreme cross on the right-hand side is taken as the reference point, and the appropriate score value range below this point is taken.

The scores are added together and the sum converted into a five-point "morningness-eveningness" scale as follows:

	Score
Definitely morning type	70–86
Moderately morning type	59–69
Intermediate	42–58
Moderately evening type	31–41
Definitely evening type	16–30

1. Considering only your own "feeling best" rhythm, at what time would you get up if you were entirely free to plan your day?

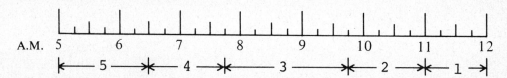

2. Considering only your own "feeling best" rhythm, at what time would you go to bed if you were entirely free to plan your evening?

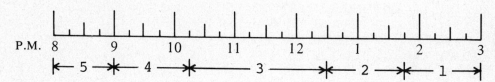

3. Is there a specific time at which you have to get up in the morning? To what extent are you dependent on being awakened by an alarm clock?

   not at all dependent _____ 4

   slightly dependent _____ 3

   fairly dependent _____ 2

   very dependent _____ 1

4. Assuming adequate environmental conditions, how easy do you find getting up in the morning?

   not at all easy _____ 1

   not very easy _____ 2

Key for Horne-Östberg Questionnaire

fairly easy _____ 3

very easy _____ 4

5. How alert do you feel during the first half-hour after waking up in the morning?

not at all alert _____ 1

slightly alert _____ 2

fairly alert _____ 3

very alert _____ 4

6. How is your appetite during the first half-hour after waking up in the morning?

very poor _____ 1

fairly poor _____ 2

fairly good _____ 3

very good _____ 4

7. During the first half-hour after waking in the morning, how tired do you feel?

very tired _____ 1

fairly tired _____ 2

fairly refreshed _____ 3

very refreshed _____ 4

8. When you have no commitments the next day, at what time do you go to bed compared to your usual bedtime?

seldom or never later _____ 4

less than 1 hour later _____ 3

1–2 hours later _____ 2

more than 2 hours later _____ 1

9. You have decided to engage in some physical exercise. A friend suggests that you do this 1 hour twice a week, and the best time for him is between 7 and 8 A.M. Bearing in mind nothing else but your "feeling best" rhythm, how do you think you would perform?

would be in good form _____ 4

would be in reasonable form _____ 3

would find it difficult _____ 2

would find it very difficult _____ 1

10. At what time in the evening do you feel tired and, as a result, in need of sleep?

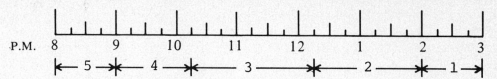

11. You wish to be at your peak performance for a 2 hour test that you know is going to be mentally exhausting. You are entirely free to plan your day. Considering only your own "feeling best" rhythm, which ONE of the four testing times would you choose?

    8:00–10:00 A.M.   _____ 6

    11:00–1:00 P.M.   _____ 4

    3:00–5:00 P.M.   _____ 2

    7:00–9:00 P.M.   _____ 0

12. If you went to bed at 11:00 P.M., at what level of tiredness would you be?

    not at all tired   _____ 0

    a little tired   _____ 2

    fairly tired   _____ 3

    very tired   _____ 5

13. For some reason you have gone to bed several hours later than usual, but there is no need to get up at any particular time the next morning. Which ONE of the following events are you most likely to experience.

    will wake up at usual time and will NOT fall asleep again   _____ 4

    will wake up at usual time and will doze thereafter   _____ 3

    will wake up at usual time but will fall asleep again   _____ 2

    will not wake up until later than usual   _____ 1

14. One night you have to remain awake between 4 and 6 A.M. in order to carry out a night watch. You have no commitments the next day. Which ONE of the following alternatives will suit you best?

    would NOT go to bed until watch was over   _____ 1

    would take a nap before and sleep after   _____ 2

    would take a good sleep before and nap after   _____ 3

    would take ALL sleep before watch   _____ 4

15. You have to do 2 hours of hard physical work. You are entirely free to plan your day. Considering only your own "feeling best" rhythm, which ONE of the following times would you choose?

Key for Horne-Östberg Questionnaire

8:00–10:00 A.M.	_____ 4
11:00–1:00 A.M.	_____ 3
3:00–5:00 A.M.	_____ 2
7:00–9:00 P.M.	_____ 1

16. You have decided to engage in hard physical exercise. A friend suggests that you do this for 1 hour twice a week, and the best time for him is between 10 and 11 P.M. Bearing in mind nothing else but your own "feeling best" rhythm, how well do you think you would perform?

would be in good form	_____ 1
would be in reasonable form	_____ 2
would find it difficult	_____ 3
would find it very difficult	_____ 4

17. Suppose that you can choose your own work hours. Assume that you worked a 5 hour day (including breaks), and that your job was interesting and paid by results. Which FIVE CONSECUTIVE HOURS would you select?

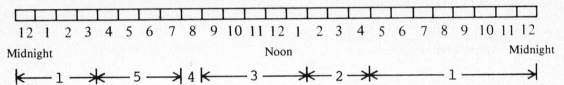

18. At what time of the day do you think that you reach your "feeling best" peak?

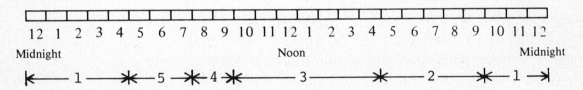

19. One hears about "morning" and "evening" types of people. Which ONE of these types do you consider yourself to be?

definitely a "morning" type	_____ 6
rather more a "morning" than an "evening" type	_____ 4
rather more an "evening" than a "morning" type	_____ 2
definitely an "evening" type	_____ 0

C. RECOMMENDED FILMS

DISTRIBUTORS

ACS   American Cancer Society (local chapter)

ALA   American Lung Association (local chapter)

BFA   BFA Educational Media, 2211 Michigan Ave., P.O. Box 1795, Santa Monica, CA 90406

CMDNJ   College of Medicine and Dentistry Library, 100 Bergman St., Newark, NJ 07103

C/M-H   Contemporary/McGraw-Hill, Princeton Rd., Hightstown, NJ 08520

EBEC   Encyclopedia Britannica Educational Corp., 425 North Michigan Ave., Chicago, IL 60611

IU   Indiana University A-V Center, Bloomington, IN 47401

JW   John Wiley and Sons, Inc., 605 Third Ave., New York, NY 10016

LL   Loma Linda University Film Library, Loma Linda, CA 92354

M-F   Milner-Fenwick, Inc., 3800 Liberty Heights Ave., Baltimore, MD 21215

MFR   Modern Film Rental, 2323 New Hyde Park Rd., New Hyde Park, NY 11040

NMAVC   National Medical A-V Center, National Library of Medicine, U. S. Public Health Service, Atlanta, GA 30333

PSU   Pennsylvania State University, Audio Visual Aids Library, University Park, PA 16802

SU   Syracuse University Film Rental Center, 1455 East Colvin St., Syracuse, NY 13210

UC   University of California Extension Media Center, Berkeley, CA 94720

UWP   University of Washington Press, Seattle, WA 98105

Blood

The Embattled Cell, 16 mm, 21 min, sd, color, ACS
    A truly excellent biomedical film dealing with the defenses of the body, emphysema, and cancer. Time lapse and phase contrast cinematography show leukocyte action in destroying cancer cells.

Genetics of Transplantation, 16 mm, 19 min, sd, color, M-F
    Description of the major histocompatibility antigens in man; demonstrates blood and tissue typing procedures; highlights a kidney transplant.

Rh—The Disease and Its Conquest, 16 mm, 18 min, sd, color, M-F
    History of the discovery of Rh disease, its physiology, and its mode of inheritance; shows actual procedures for treatment of affected babies.

Sickle Cell Anemia: Suspicion and Diagnosis in Infants and Children, ¾" videocassette, 20 min, sd, color, CMDNJ
    An excellent videotape showing the clinical diagnosis of sickle cell anemia with many actual patients. Tasteful and informative.

Sickle Cell Anemia: Management, ¾" videocasette, 14 min, sd, color, CMDNJ
  Specific therapeutic measures for the treatment of sickle cell anemia. Many clinical examples.

The White Blood Cells, 16 mm, 12 min, sd, color, IU, PSU, SU, UC
  Dramatic microcinematography shows the action of several types of leukocytes against one tubercle bacillus.

The Work of the Blood, 16 mm, 13 min, sd, color, EBEC, IU, SU, UC
  An excellent film showing techniques of blood analysis, blood cell types and their functions in gas exchange, wound healing, nutrient and waste transport, and defense of the body.

Heart

Heart Disease: Its Major Causes, 16 mm, 13 min, sd, color, SU
  A good introduction to heart disease, contrasting normal and diseased patients; heart sounds, roentgenograms, EKGs, and prosected human hearts.

Roentgenogram Anatomy of the Normal Heart, 16 mm, 17 min, sd, B&W, NMAVC
  Illustrates the flow of blood through the heart and coronary arteries by alternate use of cinefluorometric recordings and animation.

W. Harvey and the Circulation of the Blood, 16 mm, 40 min, sd, color, NMAVC
  Personal biography; contemporary theories of circulation, and actual reconstruction of the experiments leading to Harvey's hypothesis.

Permanent Transvenous Pacing of the Heart, 16 mm, 14 min, sd, color, UC
  Illustrates implantation of pacemaker; shows proper position of electrode, monitoring, measurement of stimulus threshold, etc.

The Work of the Heart, 16 mm, 19 min, sd, color, EBEC, IU, SU, UC
  This excellent film combines a variety of elements, including heart surgery, a view of the mitral valve *in situ* in a living patient, heart models, and artistic representations, to give a better understanding of the heart.

Respiration

Battle to Breathe, 16 mm, 25 min, sd, color, ALA
  A film on emphysema using three actual case histories. Interviews with patients and shots of their daily lives give viewers a feeling of what it is like to have the disease.

Diagnostic Tests for Bronchitis and Emphysema, 16 mm, 7 min, sd, color, CMDNJ
  This film gives a brief review of the apparatus and procedures used to test for obstructive lung disease.

Emphysema, the Facts, 16 mm, 13 min, sd, color, ALA
   An informal but informative approach to the question of smoking and emphysema.

Is It Worth Your Life?, 16 mm, 23 min, sd, color, ALA
   A fine presentation of respiratory function and the effect of cigarette smoking on the lungs.

Respiration in Man, 16 mm, 26 min, sd, color, EBEC, UC
   A wide variety of techniques is used to let the students see as much as possible of the structure and function of the respiratory system.

Roentgenogram Anatomy of the Human Thorax, 16 mm, 35 min, sd, B&W, NMAVC
   Illustrates lung development, anatomy of the bronchial "tree," the pulmonary vascular system, mediastinum, pleura, and the diaphragm through the use of cinefluorometric recording.

Smoking/Emphysema: A Fight for Breath, 16 mm, 11 min, sd, color, C/M-H
   An effective explanation of the likely connection between smoking and emphysema, dramatizing the developmental stages of the disease.

Spirometry, 16 mm, 27 min, sd, color, ALA
   Discusses the physiology of emphysema and demonstrates in detail the use of spirometry in diagnosing the extent of the disease. Two actual case histories are presented.

Gas Laws

Boyle's Law, super 8 mm film loop (Ealing #80-3387/1), 4 min, color, BFA
   Several simple laboratory experiments on Boyle's law are shown. Data are provided for student use.

Thermal Expansion of Gases, super 8 mm film loop (Ealing #80-3312/1), 4 min, color, BFA
   Several simple laboratory experiments on Charles' law are shown. Data are provided for student use.

Metabolism and Digestion

The Alimentary Tract, 16 mm, 11 min, sd, B&W, PSU, UC
   Live photography showing action of the esophagus, stomach, small intestine, and colon as food moves through the gastrointestinal tract.

Digestion: Part I. Mechanical, 16 mm, 17 min, sd, color, IU, PSU, SU
   Step-by-step presentation of the mechanics of digestion: mastication, peristalsis, segmentation, and absorption by blood.

Digestion: Part II. Chemical, 16mm, 17 min, sd, B&W, IU, PSU, SU
   Secretion and action on carbohydrates, proteins, and fats, of saliva; gastric, pancreatic, and intestinal juices; and bile.

Fiberscope Endoscopy of the Upper Digestive Tract, 16 mm, 14 min, sd, color, NMAVC
   Principles and use of fiber optics in filming the esophagus, stomach, and duodenum.

The Human Body: Chemistry of Digestion, 16 mm, 15 min, sd, color, MFR, SU
   Animation, X-ray footage, and laboratory demonstrations illustrate major steps in the digestion of carbohydrates, fats, and proteins.

Roentgenogram Anatomy of the Normal Alimentary Tract, 16 mm, 30 min, sd, B&W, NMAVC
   Illustrates the normal alimentary canal, from the swallowing mechanisms to movements of the colon, through alternate use of cinefluorometric recording and animation.

Nerve and Sensory Processes

Ears and Hearing, 16 mm, 22 min, sd, color, EBEC, UC
   Shows structure and function of human ear; demonstrates process by which sound is converted to electrochemical energy; includes closeups of bone operations on middle ear.

Eyes and Seeing, 16 mm, 22 min, sd, color, EBEC, UC
   Provides an understanding of the eye and its components; excellent pathologic macro- and microcinematography identifies the parts of the eye; experiments dealing with field of vision.

Human Body: The Brain, 16 mm, 17 min, sd, color, MFR, SU
   Laboratory demonstrations, X-ray footage, actual specimens, and animation are used to explain our present understanding of the brain.

The Nerve Impulse, 16 mm, 22 min, sd, color, EBEC, IU, UC
   Excellent reconstructions of classical experiments leading to present understanding of the nerve impulse (Galvani, DuBois-Reymond, Helmholtz, Bernstein, Loewi, Hodgkin-Huxley); includes oscilloscopic study and photo- and electron micrography.

Reflexes, 16 mm, 20 mins, silent, B&W, PSU
   Animated reflex arc; proprioceptive versus extroceptive reflexes; thorough clinical examination of standard human reflexes.

The Technique of Intracellular Perfusion of Squid Giant Axon, 16 mm, 10 min, sd, color, NMAVC
   A step-by-step demonstration of the preparation and study of the squid axon.

Reproductive Processes

Biography of the Unborn, 16 mm, 17 min, sd, B&W, PSU, UC
   Detailed development of the human embryo from time of conception through birth as shown by photomicrography and animation.

Ovulation and Egg Transport in Mammals, 16 mm, 20 min, color, sd, UWP
   This film illustrates ovulation in the living animal by showing the mechanism whereby the ovulated egg is transported from the surface of the ovary into the oviduct and to the site of fertilization within the oviduct.

Prenatal Diagnosis by Amniocentesis, 16 mm, 26 min, sd, color M-F
   Two genetic counseling interviews are presented in which amniocentesis is recommended. An actual procedure is shown.

Sperm Maturation in the Male Reproductive Tract: Development of Motility, 16 mm, 14 min, color, sd, UWP
   This film shows the gradual attainment of motility as spermatozoa pass from the seminiferous tubules through the ductuli efferentes, epididymis, and finally enter the ductus deferens.

Skeletal System

Human Body: The Skeleton, 16 mm, 11 min, sd, color, MFR
   Cinefluorographic studies showing movement of the skeleton and how it supports and protects the body.

Roentgenogram Anatomy of Normal Joints and Bones, 16 mm, 20 min, sd, B&W, NMAVC
   Cinefluorographic studies of the characteristic movements of the different types of joints.

Roentgenogram Anatomy of the Normal Vertebral Column, 16, 20 min, sd, color, NMAVC
   Illustrates the anatomy and function of the vertebral column through the use of models, radiographs, and cineradiographs; discusses spinal development and normal curvature at rest and in motion.

The Spinal Column, 16 mm, 11 mins, sd, color, EBEC
   Supplements live action scenes with X-ray and stop-action photography, animation, and graphic close-ups; provides a detailed study of the structure and function of the spinal column.

Muscles

Microelectrodes in Muscle, 16 mm, 19 min, sd, color, JW
   Shows measurement of transmembrane potential changes and how external ionic composition affects the electrical responses of an excitable cell.

Muscular Contraction Under the Microscope, 16 mm, 27 min, sd, color, UC
> Illustrates changes in striation, failure of highly stretched muscles to contract, and "local activation" phenomena occurring during contraction.

The Muscle Spindle, 16 mm, 19 min, sd, color, JW
> Vividly illustrates the action of the muscle spindle, alternately using photomicrography and animation.

Muscles: Chemistry of Contraction, 16 mm, 15 min, sd, color, EBEC, SU
> An excellent presentation of muscle structural proteins and the effect of ATP and calcium ions on their function, using laboratory demonstrations, molecular models, and photomicrography; the sliding filament theory is presented.

Muscles: Dynamics of Contraction, 16 mm, 21 min, sd, color, EBEC, SU
> The three types of muscles are shown and their contraction is explained in terms of the properties of muscle fibers, fibrils, and the two types of filaments.

Muscles: Electrical Activity of Contraction, 16 mm, 19 min, sd, color, EBEC, SU
> Experiments with microelectrodes demonstrate how an electrical event procedes muscle contraction by a very brief interval; this stimulus is shown to activate an "all-or-none" response.

Physical Diagnosis: Examination of the Musculo-skeletal System, 16 mm, 25 min, sd, color, IU
> An excellent film showing detailed examination of a healthy man by a physician; some of the more commonly found musculo-skeletal abnormalities are also shown.

Work vs. Load in Frog Skeletal Muscle, 16 mm, 11 min, sd, color, IU
> Isolated muscle and kymograph are used to study afterloaded and loaded muscles; includes data and formulas for calculations.

Genetic Abnormalities

PKU—Preventable Mental Retardation, 16 mm, 20 min, sd, color, PSU
> This film introduces and explains PKU and follows several children's progress on a low phenylalanine diet. The importance of early detection and treatment is emphasized.

Report on Down's Syndrome, 16 mm, 18 min, sd, color, PSU
> This film provides a good understanding of both the genetic and personal manifestations of Down's Syndrome. It follows the development of two children given the avantage of a warm family life.

Endocrine Function

The Endocrine Glands, 16 mm, 11 min, sd, B&W, EBEC, LL, UC
> Controlled animal experiments together with animated drawings are used to demonstrate the functioning of the endocrine glands.

Reproductive Hormones, 16 mm, 25 min, sd, color, IU
　　Development of the human embryo; ovarian and uterine cycles; roles of the female reproductive hormones.

Stress, 16 mm, 11 min, sd, B&W, PSU, UC
　　Dr. Hans Selye discusses stress and demonstrates the experiments that have led to basic concepts.

The Kidney

Functional Anatomy of the Human Kidney, 16 mm, 31 min, sd, color, LL
　　Through the use of photomicrography and diagrams, all aspects of kidney function are clearly related to the microanatomy and untrastructure of the nephron.

The Work of the Kidneys, 16 mm, 11 min, sd, B&W, EBEC, IU, SU, UC
　　Photomicrographic record of the many functions of the renal and urinary system; unusual views of a living kidney show the selective filtering action of the glomeruli.

Homeostasis

Regulating Body Temperature, 16 mm, 22 mins, sd, color EBEC
　　The phenomenon of temperature regulation is examined with human and animal subjects; the role of the hypothalamus is investigated.